Cooperative Management

AF167450

Series editors

Constantin Zopounidis, Production Engineering and Management, Technical University of Crete, Chania, Greece
George Baourakis, Business Economics and Management, Mediterranean Agronomic Institute of Chania, Chania, Greece

More information about this series at http://www.springer.com/series/11891

Konstadinos Mattas · George Baourakis
Constantin Zopounidis
Editors

Sustainable Agriculture and Food Security

Aspects of Euro-Mediteranean Business Cooperation

 Springer

Editors
Konstadinos Mattas
Department of Agricultural Economics,
  School of Agriculture
Aristotle University of Thessaloniki
Thessaloniki
Greece

George Baourakis
Business Economics and Management
Mediterrean Agronomic Institute Chania
Chania
Greece

Constantin Zopounidis
School of Production Engineering
  and Management
Technical University of Crete
Chania
Greece

and

School of Management
Audencia Group
Nantes
France

ISSN 2364-401X
Cooperative Management
ISBN 978-3-030-08375-5
https://doi.org/10.1007/978-3-319-77122-9

ISSN 2364-4028   (electronic)

ISBN 978-3-319-77122-9   (eBook)

Printed on acid-free paper

This Springer imprint is published by the registered company Springer International Publishing AG
part of Springer Nature
The registered company address is: Gewerbestrasse 11, 6330 Cham, Switzerland

Contents

Editorial

The Cooperative Management Book Series creates a helpful framework for creative and scholarly work on cooperative management, policy, economics, organizational, financial, and marketing aspects of cooperative communities throughout the Mediterranean region and worldwide. The main objectives of this book are to advance knowledge related to collective management processes and cooperative initiatives as well as to provide theoretical background for promoting research within various sectors wherein market communities operate (agriculture, food security, real estate, insurance, and other forms). Papers appearing in this series should relate to one of these areas, should have a theoretical and/or empirical problem orientation, and should demonstrate innovation in theoretical and empirical analyses, methodologies, and applications. Further, this series encourages inter-disciplinary and cross-disciplinary research from a broad spectrum of disciplines ranging from environmental studies to business studies and food security.

The aim of this volume is to bring together researches from the agriculture and food sectors to be combined over methodological and empirical issues regarding relationships among Euro-Mediterranean countries on topics such as food policy, trade, and environmental issues. The volume also focuses on Euro-Med food relations including sustainability, marketing, trade, and policy issues. In this respect, it is critical to examine the sustainability of food sector policies under the perspective of the scarcity of natural resources.

Moreover, the degree of freedom and possible obstacles regarding trade activities between Euro-Med countries is another crucial issue, taking also into consideration the role of marketing. Proper methods will offer crucial insights into how to build up powerful tools for decision-making, particularly today that agriculture and the economy alike are affected by a volatile political, social, and economic environment and forced to undergo severe structural changes. The increase in food prices since 2007 and the world food crisis had severe adverse effects in several countries, causing macroeconomic problems (inflation, trade deficits, and fiscal pressure), increased poverty and political instability. Policymakers have acknowledged food relations as a key strategic area for Mediterranean countries, which needs to be placed at the core of Euro-Mediterranean regional cooperation.

We would like to thank the assistant editor Georgios Manthoulis and English professor Maria Verivaki for the English proofreading. We extend appreciation to the authors and referees of these chapters, and Springer Academic Publications, for their assistance in producing this book.

Konstadinos Mattas
George Baourakis
Constantin Zopounidis

Towards a More Democratic and Sustainable Food System: The Reflexive Nature of Solidarity Purchase Groups and the Migrants' Social Cooperative "Barikamà" in Rome

Daniela Bernaschi and Giacomo Crisci

Abstract Food insecurity, poverty and migration emergencies are the biggest challenges that all modern societies have to deal with. The aim of this paper is to explore the role played by civil society in dealing with these issues through a reflexive approach. Starting from addressing the multidimensional nature of food security, this deliverable comes to adopt the concept of a "democratic and sustainable food system". This change of focus allows to encompass all the complementary issues: e.g. poverty reduction, environmental sustainability, social equity and social integration. Due to the complex nature of the matter, not only a strong local and global coordination is needed but also a reflexive approach. After introducing the concept of reflexivity, this paper addresses the crucial role that civil society organizations may play in the transition towards a more democratic and sustainable food system, focusing on solidarity purchase groups in Italy. In particular, this deliverable spotlights the collaboration between solidarity purchase groups (Gruppi di Acquisto Solidale—GAS-in Italian) and the migrants' social cooperative "Barikamà" in Rome. That shows an interesting example of how civil society organizations may cope with poverty reduction, social integration and sustainability.

Keywords Democratic and sustainable food system · Civil society organizations Migrants' social cooperative

D. Bernaschi (✉)
Department of Political and Social Sciences, University of Florence and Turin,
Florence, Italy
e-mail: daniela.bernaschi@unifi.it

G. Crisci
Department of Planning and Geography, Cardiff University, Cardiff, UK

1

1   Introduction

Three relevant key points characterize this paper. Firstly, it takes into account the Post-2015 Agenda presented by the United Nations at the Sustainable Development Summit (New York, 2015). The first two SDGs to be reached by 2030 are the eradication of extreme poverty and the "Zero Hunger" program. There, the main ideas are to achieve food security and improve nutrition through sustainable agriculture. Poverty, food insecurity and sustainable agriculture are deeply interrelated phenomena.

Hence, policy strategies to achieve global food security may have to deal with a wide range of issues: climate change, poverty reduction, social equity, hunger e.g. (Arcuri et al., 2015; Marsden & Sonnino, 2012). Indeed, a key assumption in this paper is that food security is a multidimensional concept; it goes beyond mere food availability and entails different kinds of deprivation. With this in mind, the present deliverable adopts the all-encompassing concept of a democratic and sustainable food system.

Secondly, this paper discusses the link between a food system defined as democratic and sustainable and the "reflexive" approach as a way to better coordinate local and global issues (Duncan, 2015; Edwards et al., 2002; Marsden, 2013; Voss & Kemp, 2005, 2006; Wolff, 2006). Reflexive governance is characterized by a multi-stakeholder structure, interactive participation and social learning. Despite their various declinations, the academic debates on reflexive governance agree that a broader social participation, discussion and sharing of knowledge are crucial to solve social problems. Hence, this paper illustrates the emerging contribution offered by the reflexive governance in addressing new social issues.

Thirdly, it explores the role of civil society organizations in their attempt to attain a more democratic and sustainable food system. The article analyses the three specific contributions of solidarity purchase groups (Gruppi di Acquisto Solidale—GAS-in Italian) in terms of: changing consumers' behavior and improving nutrition, supporting local development and enhancing social integration. Then, the collaboration between the GAS Movement and Barikamà will be discussed. Barikamà is a cooperative founded by six migrants from Sub-Saharan Africa in 2014. Its experience demonstrates how civil society organizations can positively improve social integration and contribute to the struggle against poverty, realizing a more democratic and sustainable food system.

2   The Unsustainable Global Food System: The Paradox of "Scarcity in Abundance"

"We live in a world of unprecedented opulence (...) People live much longer, on an average, than ever before. (...) And yet we also live in a world with remarkable deprivation, destitution and oppression. (...) Overcoming these problems is a central part of the exercise of development" (Sen, 1999, Preface).

The etymological meaning of "development", dating back to the mid-18th and late 19th century, is "an unfolding" (1756), "bringing out the latent possibilities" (1885). In other words, it implies the removal of any obstacle that prevents and limits the free and independent development of human beings to lead the kind of life they value and have reason to value (Sen, 2000a). Broadly, it refers to the Aristotelian concept of "eudaimonia", whose meaning is well captured by the idea of "human flourishing".

At the core of the Sustainable Development Goals (SDGs) set by the United Nations in September 2015, there is the official commitment to overcome the main forms of deprivation that prevent people from flourishing. The new global development framework articulated in 17 main goals to be reached by 2030 endorses a more sustainable and inclusive development.

As Daily et al. (1998) stated, there are two main criteria to assess mankind's capacity to feed itself: (i) whether individuals can have a secure access to basic nutritional needs; and (ii) how global food production is sustainable. Considering these criteria, we are experiencing a deep flaw in the global food system. Indeed, about 795 million people in the world are undernourished (FAO, 2015), more than 1.9 billion adults are overweight and among these 600 million people are obese, and the numbers are growing (WHO, 2014).

Furthermore, a global food system is increasingly vulnerable and failing from a sustainability point of view. According to the IPCC Report (2014), 24% of greenhouse gas emissions in the world derive from deforestation and methane produced by livestock, paddy fields and the use of fertilizers in agriculture.

Agricultural input intensification leads to soil degradation, losses in biodiversity and overuse and pollution of water (Duncan, 2015; Godfray et al., 2010; Hazell & Wood, 2008). Moreover, according to FAO (2013a, b), every person wastes 350 kg of food per year, and this waste itself leads to a massive production of greenhouse gasses only just inferior to the emissions of USA and China.

In conclusion, the current global food system seems to be environmentally unsustainable and characterized by "scarcity in abundance", as Campiglio and Rovati (2009) defined this paradox: on the one hand, there is overconsumption and waste production; on the other hand, 795 million people are undernourished. In other words, the current food system leads to inequalities in the access and distribution of resources.

3   From the "Multidimensional Nature" of Food Security to a More Democratic and Sustainable Food System

To explore the case study we are going to present, it is fundamental to understand how the concept of food security has changed in the last forty years. Since the first definition given by the World Food Conference (1974), where food security was defined merely in terms of food supply, the concept has evolved consistently and is still at the center of the academic debate. Despite the crises that have occurred over the years, the global availability of food is sufficient to ensure almost 2800 kcal/ person/day (FAO, 2006), an amount that exceeds our needs. Looking at the figure related to FAO's data (2013a, b), we can also notice an unequal distribution among countries: for instance, North America, France, Portugal, Italy, Greece, Romania and Turkey are marked by more than 3,276 kcal per capita per day; whereas Bolivia, Ecuador, Kenya, Ethiopia are marked by less than 2,454 kcal.

Over the last decades, a new definition of food security has been developed in order to address its multidimensional nature. A great contribution to the elaboration of a more comprehensive approach to food security was offered by Amartya Sen at the beginning of 1980s.

Sen introduced an "entitlements-based analysis" in order to address famine and hunger. In his famous work "*Poverty and Famines*" (1981), the focus of the analysis shifted from food availability at the national level (Fig. 1) to people's entitlements. Hunger resulted as an economic issue, rather than a humanitarian one.

Whether we are unable to buy enough food to satisfy our hunger, then we are destined to suffer from it (Sen, 2000b).

In a market economy, entitlements depend inter alia on the resources and endowments available: labour force, land and the possession of production factors. These said factors can be directly used or sold in the market. The entitlements are related to the market opportunities for what we sell, and the prices and availability of food and other goods that we want to buy. Getting enough food to eat or otherwise be forced to go hungry depends on many aspects, such as the endowments and the conditions of production and exchange. The totality of these factors determines the type of entitlements.

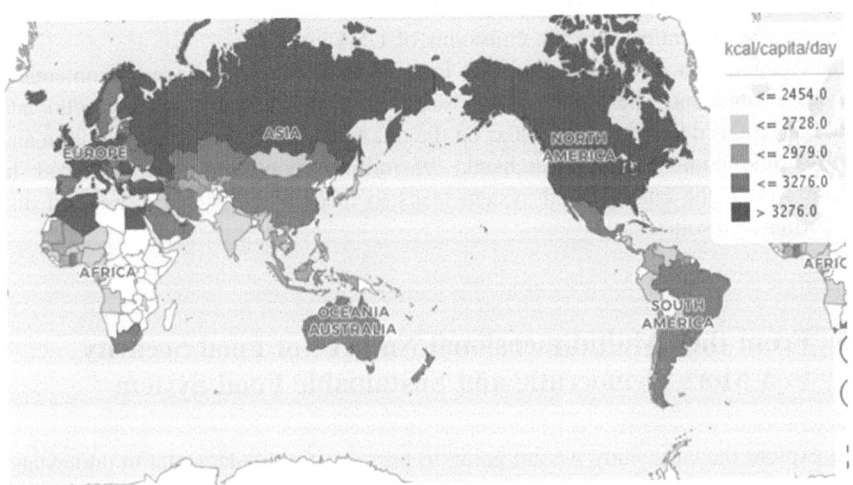

Fig. 1 Global food availability. *Source* FAO Statistics Division (2013b), Food Balance Sheets, Food and Agriculture Organization of the United Nations, Rome, Italy, http://www.fao.org/faostat/ en/#data/FBS/visualize, Reproduced with Permission

Dahrendorf (1989) compared the entitlements to the tickets that allow access to a desired place. Through these tools, you can have access to different combinations of tangible and intangible assets that Dahrendorf called "provision". The entitlements approach highlights the institutional, political and socio-economic conditions linked to food insecurity (Burchi & De Muro, 2012).

Sen's research has affected the following definitions of food security, such as that elaborated by the World Food Summit (1996), according to which food security is the condition where: "All people, at all times, have physical and economic access to sufficient, safe and nutritious food that meets their dietary needs and food preferences for an active and healthy life".

It may be useful to mention *"Hunger and Public Action"* (1989), a joint work by Sen and Drèze, which had the merit to broaden the debate about food security by turning the attention to food utilization, and therefore to nutritional capabilities.

"The object, in this view, is not so much to provide a particular amount of food for each. Indeed, the relationship between food intake and nutritional achievement can vary greatly depending not only on features such as age, sex, pregnancy, metabolic rates, climatic conditions, and activities, but also access to complementary inputs" (Drèze & Sen, 1989: 13).

In general, food security is not achieved when all individuals get the same amount of food, because the ability to draw nourishment from it depends on multiple conversion factors related to personal characteristics (age, sex, physical and mental health) and the family, social, economic and institutional environment.

Hence, food security does not reside in the dualism "production-consumption" (Marsden & Sonnino, 2012). As shown by Drèze and Sen (1989), we must look at the use of food and complementary inputs (whether people can have access to drinking water, public health services, basic education, epidemics prevention programs) that have an impact on an individuals' health.

In Sen's words: "Hunger and undernutrition are related both to food intake and to the ability to make nutritive use of that intake. The latter is deeply affected by general health conditions and that in turn depends much on communal health care and public health provisions" (1995: 115).

The stress on the multidimensional nature of food insecurity led Sen to use terms such as "hunger" and "nutritional deprivation" and to avoid phrases like "food security", so as to remove all emphasis from "food" (Burchi & De Muro, 2012).

The need for a multidimensional approach to food security has triggered academic debates. The focus of the analysis shifted from a "production–consumption framework" to all the complementary inputs: e.g. poverty reduction, environmental sustainability issues, promotion of social equity.

The debate about access to food is enriched by many academic studies that highlight the emergence of what Sonnino (2016) defines a "new geography of food security", namely the new spatial coordinates of the problem. In fact, the compass of food insecurity is no longer rigidly fixed on the Global South and in the rural areas but it also affects the Global North (Dowler & Lambie-Mumford, 2015), where the paradox of "scarcity in abundance" (Campiglio & Rovati, 2009) is reified.

Food is available but some segments of the population are unable to access food that meets their nutritional and cultural needs (Maino et al., 2016; Shetty, 2015). This condition is often described as "food poverty" and it entails multiple deprivations (Sonnino & Hanmer, 2016).

Maslen et al. (2013: 4) argue that "food poverty is complex and multi-faceted. It is not simply about immediate hunger and how that might be alleviated. It is not just about the quantity of food that is eaten, but involves the dietary choices, the cultural norms and the physical and financial resources that affect which foods are eaten, ultimately impacting on health status".

As discussed by Sen (1995), the paradox of hunger in rich societies can be understood only if we do not focus exclusively on the income dimension.

In this sense, Lang and Barling (2012) offer a conclusive consideration. They assume that the complexity and the multidimensional nature of food security, together with the diverse perspectives that its meaning implies, claim for a broader understanding of that concept, and go on to argue that the concept itself may be limited. For this reason, they suggest giving up the expression "food security" and opting for a broader and all-encompassing definition: the "sustainable food system".

Following Lang and Barling, the current deliverable adopts the concept of "democratic and sustainable food system". In this way, the multidimensional nature of the issue will be easier to comprehend. This change of focus will also encompass poverty reduction, environmental sustainability and social integration.

4   The Reflexive Role of Civil Society Initiatives

Considering what was previously realized in this paper, it seems pivotal to follow Beck's invitation (2003) to renew the conceptual tools, in order to understand the contemporary social changes and the complexity of the matter.

Hence, it brings us to state that a more democratic and sustainable food system requires not only a coordination between local and global dimension (Duncan, 2015), but also governance arrangements which are reflexive and able to adopt flexible and prompt strategies (Voss & Kemp, 2005, 2006; Wolff, 2006; Marsden, 2013; Edwards et al., 2002).

Reflexive governance is a management strategy that does not imply a one-way problem-solving analysis (Voss & Bornemann, 2011), but an interactive and participative multi-stakeholder process, which considers different prospective analyses and strategies (Feindt, 2010; Gottschick, 2013) to reach a shared solution on the common social problems (Sonnino et al., 2014).

Studies about reflexive governance are indebted to the concept of "reflexive modernity" formulated by Giddens (1990), Beck (1992) and Lush (1994), according to whom reflexivity is "the capacity of an individual subject to direct their awareness towards themselves, reflecting upon their own practices, and constantly examining and reforming these practices in the light of incoming information" (Giddens, 1990: 38).

In reflexive governance, there are new social and political spaces, which engage public, private and civic actors and allow a common reflection on practices to be adopted in order to pursue the transition towards sustainability (Marsden, 2013) and a better understanding of social change.

The more the governance process is participative, the more it is able to reach the substantial goals of sustainability: wellbeing, social justice, respect for the environment (Stirling, 2009).

In other words, despite their various declinations, the academic debates on reflexive governance agree that a broader social participation, discussion and sharing of knowledge are crucial to solve the social problems (Marsden, 2013; Sonnino et al., 2014). The reflexive approach's essence is apparent within the civil society initiatives in their attempt to deal with the socio-economic challenges affecting modern societies.

Those initiatives draw our attention to their adaptive and responsive nature that make them able to develop innovative measures and creative responses to new forms of social vulnerability. Due to their locally based nature, civil society initiatives have a direct and deep knowledge of the social realities in which they are linked. This allows them to be more likely to capture the social transformations taking place, react promptly and adapt their strategies.

It is sharply apparent that the number of civil society initiatives have increased remarkably since the economic crisis which engulfed Europe in 2008, destroying "employment, production and human security" (Sen, 2012: 6).

The rise in poverty, social exclusion and food insecurity rates (Eurostat, 2015) has led to the flourishing of resilient civil society initiatives which work for a more democratic and sustainable food system, such as: food banks, soup kitchens, solidarity purchase groups, urban agriculture, social markets, emporiums of solidarity.

What prompts the civil society to organize itself? At the core of the community-based initiatives, there is what Beck (2003) calls "empathetic imagination", unifying and universal feeling that promotes and justifies actions and leads to open new communication and relational channels (Magatti, 2005).

However, Crespi's definition of "solidarity" seems to be more accurate with this paper's contents. According to Crespi (1994, 2013), solidarity is an expression of shared responsibility toward society, not just a selfless approach towards others.

Therefore, as Crespi (2013) stated, assuming the "constitutive sociability" of human beings, solidarity can be defined as an emancipation process through participation in community life. In other words, the constitution of a more democratic and sustainable food system requires a greater sense of responsibility toward society, caring more considerately for the world (Pulcini, 2009). Therefore, solidarity is crucial to ensure the survival of mankind and to promote inclusive development.

In the deeper civic engagement, we can see not only a desire for a broader participation, but also a social laboratory with innovative capacity and the ability of humanizing social contexts (Magatti, 2005).

Civil society ensures that the local dimension "repossesses challenges that come from the global dimension autonomously"[1] (Leonardi, 2001: 39). In fact, the social actors of civil society work to build what Beck (2003) calls "cosmopolitan society", a society which has a strong local anchoring without losing the linkages with the global dimension, a society able to promote connections instead of exclusions (Keane, 2003).

At the core of civil society initiatives (e.g. urban gardens, solidarity purchase groups and social market), there is a dense network of solidarity ties used to deal with social issues.[2]

If the top-down initiatives for local development are drawing increasing attention, the initiatives of civil society have not yet received the attention they deserve (Moulaert et al., 2005).

For this reason, the aim of the following paragraph is to increase the theoretical and practical understanding of civil society organizations. It will focus on the experience of solidarity purchase groups in Rome and their reflexive role in: changing consumers' behavior and improving nutrition; supporting local development; and enhancing social integration. Concerning the last point, social integration will be addressed, introducing the support offered by the GAS Movement to six African migrants in setting up their social cooperative called Barikamà.

5   Solidarity Purchase Groups and the Cooperative "Barikamà" in Rome: Sustainability and the Inclusion of the Socially Excluded

Solidarity purchase groups (Gruppi di Acquisto Solidale—GAS-in Italian) are grassroots networks where individuals purchase foodstuffs and other products of everyday use directly from small-medium local producers.

Why are they called "solidarity" purchase groups? Because they combine the respect for the environment with the sympathy for producers and workers to obtain a genuine sustainable and inclusive development. The crucial aspect of these groups is that the producer and consumer can establish a trusted relationship based on the shared values of solidarity and eco-friendliness. These groups aspire to rebalance the relationship between rural and urban areas, leading people to a greater and more conscious respect for the land. Sometimes, GAS's members visit local producers to learn about the sustainable agricultural practices and to help them to collect fruits

---

[1]Our translation.

[2]An example is the "Eutorto", an urban garden run by workers on redundancy payment which belongs to Eutelia Information Technology in Rome. In this case, civil society developed alternative responses to the economic crisis and a group of workers decided to remain united to cope with unemployment.

and vegetables. These groups enable the culture of sustainability, a process that should engage people and enforce their relationship with nature.

GAS are mainly informal groups and operated without any legal acknowledgement until 2007, when the Italian governmental Budget Committee released an amendment (Fifth Section) attached to the Finance Act. At a local level, several regional, provincial and municipal authorities supported the emergence and empowerment of GAS with small economic incentives (Schifani & Migliore, 2011).

Before that time, however, most of the GAS operated like single consumers or as no-profit organizations or social promotion associations (Fonte & Salvioni, 2013). Nowadays, even if they can still preserve their flexible legal form, the formal recognition from the authorities has been relevant to classify them as a non-commercial activity.

The GAS movement is organized in subgroups that act independently. According to Retegas (1999), the activity of every single group is directed to create a space for an ethical and sustainable food consumption which puts people together rather than divide them.

A study led by the Coldiretti Farmers Union (2014) shows that from 2008 the movement has increased by 400% in Italy, and the number of Italian citizens that regularly arrange collective orders has reached 2.7 million.

The diffusion of these new forms of solidarity and social economies arises from a new reflexive capacity in the consumers (Giddens, 1990), who reflect on their own actions and modify them according to new incoming information. Furthermore, the unprecedented rise of these types of economies can be seen as a specific form of social self-defense (Guthman, 2007), since they are a response to the recession and to social retrenchments.

This paper aims to spotlight how GAS can contribute towards a more sustainable and democratic food system. In this regard, it will address GAS's contribution in: reshaping consumers' behavior and improving nutrition; supporting local development; and enhancing social integration.

As for the first point, local food networks are the alternative systems through which individuals can have economic and physical access to food for a fair price (Goodman & DuPuis, 2002; Renting et al., 2003). According to Crisci and Fonte (2014), GAS's members save up to 72.5% weekly, more than what they spend when buying organic food through other sales channels. Since the vegetarian purchase in GAS is less expensive than the omnivorous one, the reduction of meat consumption renders GAS more competitive and improves the sustainability of agrifood systems (Ibidem, 2014).

Hence, GAS can improve the economical and physical access to fresh and organic food. In addition, the trend in consumption and lifestyle changes consistently after joining a GAS: people consume more organic, seasonal and local food, limit the consumption of meat and start to produce home-made food (Grasseni et al., 2013).

The solidarity purchase groups can also reinforce local development. In fact, their main goal is to support small-medium organic producers, who cannot survive in the large-scale trade because they are not strong enough to compete in the mainstream food system (Retegas, 1999). The fair price promoted by GAS is the result of an agreement between consumers and producers.

Considering the difficulties sometimes faced by small enterprises, GAS's members set up pre-financing systems where consumers pay for crops in advance and share part of the risk taken by the farmer (Savioli et al., 2015). This gives the producers disposable cash to use for further cultivation and the certainty to sell the product at the pre-arranged price (Grasseni et al., 2013). Supporting local development leads to the boost of territorial heritage and biodiversity.

Concerning the social integration promoted by GAS, it has recently become a central element in the alternative food network, thanks in particular to the engagement of the SOS Rosarno organization (Oliveri, 2015). This organization sets up alternative economic strategies and new social alliances in order to overcome small producers' impoverishment, migrant farmworkers' exploitation and racial discrimination in many Italian regions.

SOS Rosarno deliveres oranges, lemons and other products to the GAS spread in Italy and still represents the main solution to racial exploitation in Southern Italy. Concerning the role that GAS may play in promoting social integration, it may be interesting to discuss the relationship between the Roman GAS Movement and Barikamà, a cooperative set up by six migrants in 2014.

Suleman, Aboubakar, Sialiki, Modibo, Ismael and Moussa are six young men from Mali. They crossed the desert by foot in one year; during that excruciating journey, many of their friends and travel companions did not survive. After that, they arrived in Libya and then they travelled by boat to finally land on the Italian coast.

Because of its geographical position Italy, like several Mediterranean countries, has become an unavoidable passage to escape from the migrants' original countries. Monzini (2007) describes the process through which these people become part of an illegal system in Southern Italy. When they arrive in Italy, migrants are exploited by illegal organizations: they are forced to pay for some services and are subjected to rough treatment.

Barikama's members (Fig. 2) were exploited in the tomato and citrus production in South Italy, and in particular they started working in the tomato agri-business in Puglia. There, migrants work from 8 to 12 h a day, with wages that range between 20 and 25 Euros per day, in working environments with little or no safety measures at all (Bernaschi, 2013).

The Amnesty International Report (2014) states that migrant workers in Italy are victims of 'severe exploitation' in the agricultural sector. The debilitating working conditions lead to human rights violations and deprivation of basic capabilities, firstly health: these people suffer from diseases related to the absence of safety measures, malnourishment and undernourishment. Moreover, an unsecure job determines housing conditions that are degrading from a socio-sanitary point of view, like makeshift accommodations and overcrowding (Pugliese et al., 2012; Bernaschi, 2013).

Fig. 2 Barikamà and GAS's members. *Source* Barikamà's website http://barikama.altervista.org/foto-e-video/, Reproduced with Creative Commons Permission

As Brovia (2008) reported, more than 12,000 migrants work in the seasonal harvest of tomato crop in Northern Puglia. Most of them are employed illegally through the so-called "caporalato" system that is a gangmaster system.[3]

In 2006, after a decisive social pressure related to the living conditions of migrant workers, several initiatives were started to provide an alternative response to exploitation. One of these initiatives was promoted by the SOS Rosarno organization, which put a group of six migrants (soon to become Barikamà's associates) in contact with activists able to introduce them into the local food system in Central Italy.

In January 2010, after a violent riot against exploitation and racism occurred in Rosarno (Calabria region, Southern Italy), the six migrants moved to Rome. There, they came across members of the Ex Snia community center and established contact with the GAS movement (Diara et al., 2015).

Thanks to that encounter, Barikamà became a social promotion association in order to gain legal acknowledgement. In the Bambarà language, Barikamà means "resilience", showing the adaptive and responsive nature of this civil society initiative. The Barikamà experience touches several points related to social inclusion: learning a new language and a new job; community building; inclusion in the workforce; crowdfunding; and the inclusion of Italian workers in the social cooperative "Barikamà".

---

[3]The Ethical Trading Initiative Report (2015), an alliance formed by companies and NGOs committed to respecting the rights of the workers in the world, denounced the massive exploitation of migrants in Southern Italy's tomato industry and the involvement of the Mafia. As a consequence, some Northern European countries have decided to boycott Italian tomatoes.

Fig. 3 Barikamà's yoghurt: production and distribution. Source: Barikamà's website http://
barikama.altervista.org/foto-e-video/, Reproduced with Creative Commons Permission

In the beginning, they relied on the African food culture (mainly based on family
livestock production) and on the knowledge gained both in their countries as well as
in Italy, so they decided to produce yogurt to be sold to farmers' markets and to
GAS members.

GAS members gave them the opportunity to learn how to produce it with the
assistance of a local organic dairy producer who trained them. The response of the
consumers was mostly based on the willingness of several activists to test the
product and stay in contact with these migrants, so as to facilitate their integration
and develop a marketable product. Through this process, Barikamà's migrants
learned new skills: dairy production, marketing and distribution (Fig. 3).

They presented the project "Barikamà" in the farmers' market and during the
GAS meetings and this allowed the migrants to work in a team and to learn Italian.

Then, thanks to the support of "Casale di Martignano", a certified organic farm,
they were able to transform their informal job into an enterprise that respected all
the sanitary requirements. In 2014, they started a social cooperative "type B"[4]
aiming to integrate disadvantaged people into the labour market.

According to GAS's set of principles, farming methods should always be
organic. The migrants started to produce organic yogurt even without any formal
certification; firstly, because their customers knew that labeling is expensive for
small producers and could make them less competitive (Pimbert et al., 2006) and
secondly, because the milk they used to make the yogurt was organic.

Today, the yogurt they sell shows an organic certification since it is produced in
a fully organic dairy farm.

In 2014, Barikamà participated in a public regional competition launched by the
regional government concerning the solidarity economy and got an assisted loan of
20,000 Euros. However, since that money had to be spent before the government

---

[4]In social co-operatives type B at least 30% of the members must be from disadvantaged groups:
people with physical and mental disability, drug and alcohol addiction; ex-convicts e.g.

started the loan, the cooperative launched a crowdfunding initiative and gathered 26,000 Euros in a few months.

GAS's support was crucial, since they loaned Barikamà a considerable amount of money and accepted to be refunded in food products during the following years. The competition allowed Barikamà to buy all the equipment needed to improve the quality of their yogurt and to do home deliveries, for which they needed electric bicycles and new refrigerators.

Therefore, Barikamà also started to work as a bike company with a project called "Pedalatte Bio", which literally means "organic milk cycling" to delivery milk, yogurt and other organic dairy products in Rome. They decided to use bicycles as a more environmentally-friendly transportation method, harmonising with Barikamà's meaning of resilience.

In addition, the crowdfunding initiative not only allowed the migrants to gather the money needed but improved their network of potential partners and customers. Consequently, they developed enough relational skills to maintain and cultivate relationships with other associations and social entities.

Hence, Barikamà is the final output of a dense network of ties between associations and different civil society initiatives that deal with poverty, working exploitation and discrimination, all complying with respect for the environment.

Moreover, thanks to the skills acquired in Barikamà, many members (who only work part-time in the cooperative) found other jobs in restaurants and farms around Rome. Many of them were able to earn decent incomes in just a few years and even invest in their business.

Another relevant point of this scaling-up experience is that today the social cooperative "Barikamà" also allows Italian people affected by disabilities to work in its staff. Barikamà has decided to employ two young Italians who suffer from Asperger's syndrome (a form of autism), driven by the empathy towards them.

Indeed, when the migrants arrived to Italy they were not able to communicate due to their language and felt isolated. However, thanks to the support offered by the solidarity networks, Barikamà has started to rise as a relevant link in a social promotion chain, fostering the inclusion of the socially excluded.

Hence, the Italians who have Asperger's syndrome, dealing with the cooperative's website and home deliveries, can earn money, gain new knowledge and improve their personal skills.

Barikamà represents an outstanding demonstration of how the integrated and reflexive civil society initiatives can positively affect the local community, by integrating migrants and other disadvantaged people.

Recently, the Restaurant "Grandma Bistrot", the cooperative "Barikamà" and the organic farm "Casale di Martignano" decided to participate together in a public competition to run an organic local café. Thanks to their perseverance and appeal to the institutional initiatives, they won the competition and are about to open the café in a former Mafia store requisitioned by the authorities. The organic café is located in Parco Nemorense, a town park in Rome whose conservation is managed by Barikamà together with the other business partners.

6   Findings

Following what was said in the previous paragraph, the Barikamà project is scaling up and attempts to combine several development dimensions.

Indeed, the collaboration between the GAS movement and Barikamà in Rome shows an interesting example of how civil society organizations may cope with poverty reduction, social integration, sustainability and local development.

GAS and Barikamà are what Isin (2009) called an "act of citizenship", namely a social act that facilitates the transition from the status of "citizen, foreigner, outsider" to that of a "citizen activist", who demands his rights and exercises his capability for voice (Bonvin et al., 2013).

This study has three aims. Firstly, it tries to understand whether and how Barikamà contributes to reducing poverty. Barikamà as a project is based on a deep synergy and interdependence between different types of civil society actors moved by solidarity in terms of responsibility towards society.

Thanks to these interconnected relationships and thanks to a reflexive approach adopted, Barikamà started to rise as a new entity. The cooperative provides subsistence income to the migrants. In addition, through the solidarity networks the migrants found other supplementary jobs.

Secondly, it tries to comprehend the role played by Barikamà in contributing to the social inclusion of migrants and disadvantaged people through environmentally friendly projects. The primary purpose of Barikamà is to allow a gradual inclusion of migrants by developing several working skills. Indeed, Barikamà members started working and developing skills related to organic yoghurt production. Consequently, Barikamà represents a means through which migrants and the two young Italians with disabilities can improve their professional skills and be employed.

Thirdly, it tries to assess how reflexivity works in the relationships between Barikamà and the GAS movement as discussion and sharing of knowledges. Starting from their common attempt to support a more sustainable food system, Barikamà decided to produce organic food to be delivered on bicycles. Furthermore, the cooperative fosters local development: e.g. the conservation of Parco Nemorense, running the confiscated Mafia store.

Barikamà as a "development project" allows us to understand how the synergies between the different social initiatives are crucial and how civil society is creative in dealing with new and complex issues. However, reflecting on the migrants' social cooperative, the collaboration with the public institutions seems pivotal as it has allowed the early project to get expanded and strengthened.

In other words, civil society organizations may play a strategic role in the transition towards a more democratic and sustainable food system. Nevertheless, as Leonardi (2001) stated, they need to establish a dialectical relationship with the institutions to reinforce their role and to become more important. At the same time, public institutions should underpin and be inspired by the innovative projects led by those initiatives that are strongly linked to community life. Hence, this paper provides evidence of why it is opportune to foster virtuous synergies between civil society initiatives and the public institutions.

7 Conclusion

This paper, starting from a holistic vision of the world in terms of interconnected and interdependent phenomena (Capra & Luisi, 2014), suggests analyzing food security in its multidimensional dimension. An all-encompassing comprehension of the problem is possible only by deleting the dichotomy "production-consumption" and by broadening the assessment towards: e.g. poverty reduction, environmental issues and social integration. With this in mind, the concept of a "democratic and sustainable food system" has been considered as more suitable to address the topic.

Moreover, a strong local and global coordination as well as a reflexive approach is required in order to analyze the matter extensively. This paper assesses the growing civic engagement around food through its reflexive nature of promoting a wide social participation, debate and cooperation for the solution of shared social problems.

This deliverable aims to deal with those civil society initiatives that contribute to a more democratic and sustainable food system. On account of this, it focuses on the solidarity purchase groups and their relationships with the migrants' cooperative "Barikamà" in Rome. That experience allows us to spotlight three relevant considerations:

Firstly, civil society initiatives are highlighted. Thanks to a deep knowledge of the social realities to which they are linked, they are able to develop creative, dynamic and reflexive measures to address poverty, social integration and sustainability.

Secondly, those initiatives represent a form in which individuals take care of the world as a form of responsibility towards society, fostering relationships instead of exclusions.

Lastly, civil society initiatives based on dense networks of solidarity ties may play a strategic role in the transition towards a more democratic and sustainable food system. Nevertheless, in order to matter they need to come out of the niche dimension. This can be done by building an interconnected collaboration with public institutions.

This paper provides evidence of why social issues to be addressed require more than just a fertile civic ground. Indeed, an involvement of public institutions seems necessary in supporting the civic society initiatives (e.g. the legal acknowledgement received by the GAS movement in 2007 or the financial support received by the Barikamà cooperative) and in assuming an active role in tackling new and complex social issues.

References

Amnesty International. (2014). *Exploited labour two years on: the "Rosarno Law" fails to protect migrants exploited in the agricultural sector in Italy*. London, UK: Amnesty International Publication.

Arcuri, S., Brunori, G., Bartolini, F., & Galli, F. (2015, June). "La sicurezza alimentare come diritto: per un approccio sistemico." *Agriregionieuropa*, Year 11 (Vol. 41).

Beck, U. (1992). *Risk society: Towards a new modernity*. London, UK: SAGE.

Beck, U. (2003). *La società cosmopolita. Prospettive dell'epoca postnazionale*. Bologna, Italia: Il Mulino.

Bernaschi, D. (2013). *Analisi delle condizioni di vita e di lavoro dei migranti nell'area del Casertano. Tra sfruttamento lavorativo e privazioni delle basic capabilities*. Master's thesis in Human Development Economics.

Bonvin, J. M., Dif-Pradalier, M., & Moachon, E. (2013). *A capability approach to restructuring processes, lessons from a Swiss and a French case study*. Lausanne, Switzerland: CESCAP, University of Applied Sciences Western Switzerland.

Brovia, C. (2008). In the iron grip of the Caporali: seasonal workers on tomato farms in Apulia; Sous La Férule Des Caporali. Les Saisonniers De La Tomate Dans Les Pouilles. *Etudes Rurales, 182*(2), 153–166.

Burchi, F., De Muro, P. (2012). A human development and capability approach to food security: Conceptual framework and informational basis. United Nations Development Programme, WP 2012-009, February 2012.

Campiglio, L., & Rovati, G. (2009). *La povertà alimentare in Italia: Prima indagine quantitativa e qualitativa*. Milano, Italia: Guerini e associati.

Capra, F., & Luisi, P. L. (2014). *The system view of life*. Cambridge, UK: Cambridge University Press.

Coldiretti. (2014). Crisi: Coldiretti, spesa di gruppo per 2, 7 mln italiani (+400%). Retrieved June 21, 2017 from http://www.coldiretti.it/News/Pagine/672–%E2%80%93-12-Ottobre-2014.aspx.

Crespi, F. (1994). *Imparare ad esistere. Nuovi fondamenti della solidarietà sociale*. Roma, Italia: Donzelli.

Crespi, F. (2013). *Esistenza-come-realtà. Contro il predominio dell'economia*. Salerno, Italia: Orthotes.

Crisci, G., & Fonte, M. (2014). L'accesso al bio nella transizione verso la sostenibilità dei sistemi agro-alimentari. *Agriregionieuropa, year 10*(37).

Daily, G., et al. (1998). Food production: Population growth, and the environment. *Science, 281*, 12–91.

Dahrendorf, R. (1998). *Il conflitto sociale nella modernità*. Roma-Bari, Italia: Editori Laterza.

Diara, S., Crisci, G., & Fonte, M. (2015). "Barikamà: resistance through food". In *Proceedings of Second International Conference on Agriculture in an Urbanizing Society: Reconnecting Agriculture and Food Chains to Societal Needs, 14–15 September 2015*. Rome, Italy.

Dowler, E., & Lambie-Mumford, H. (2015). Introduction: Hunger, food and social policy in austerity. *Social Policy and Society, 14*(3), 411–415.

Dreze, J., & Sen, A. (1989). *Hunger and public action*. Oxford, UK: Oxford University Press.

Duncan, J. (2015). Greening global food governance. *Canadian Food Studies, 2*(2), 335–344.

Edwards, R., Ranson, S., & Strain, M. (2002). Reflexivity: Towards a theory of lifelong learning. *International Journal Lifelong Education, 21*(6), 525–536.

Ethical Trading Initiative. (2015). Due diligence in agricultural supply chains: Counteracting exploitation of migrant workers in Italian tomato production. Joint Ethical Trading Initiatives, Norway.

Eurostat. (2015). People at risk of poverty or social exclusion. Retrieved July 21, 2017 from http://ec.europa.eu/eurostat/statisticsexplained/index.php/People_at_risk_of_poverty_or_social_exclusion.

Feindt, P. (2010). Reflexive governance of global public goods: multi-level and multi-referential governance in agriculture policy. In Brousseau, E., et al. (Eds.), *Reflexive governance for public goods*. Boston, USA: MIT Press.

Food and Agriculture Organization. (1996). Rome declaration on world food security and world food summit plan of action. In *Proceedings of World Summit on Food Security, FAO*. Rome, Italy.

Food and Agriculture Organization. (2006). *World agriculture: Towards 2030/2050 Interim report prospects for food, nutrition, agriculture and major commodity groups*. Interim Report, FAO: Rome, Italy.

Food and Agriculture Organization. (2013a). *Food wastage footprint. Impacts on natural resources*. Summary Report, FAO. Rome, Italy.

Food and Agriculture Organization. (2013b). *Food balance sheets*. Statistics Division, Rome, Italy. Retrieved July 21, 2017 from http://www.fao.org/faostat/en/#data/FBS/visualize.

Food and Agriculture Organization. (2015). The state of food insecurity in the world 2015. Meeting the 2015 international hunger targets: Taking stock of uneven progress. Report, FAO. Rome, Italy.

Fonte, M., & Salvioni. (2013). Cittadinanza ecologica e consumo sostenibile: dal biologico ai Gruppi di Acquisto Solidale. In Corrado, A., & Sivini, S. (Eds.), *Cibo locale. Percorsi innovativi nelle pratiche di produzione e consumo alimentare*. Napoli, Italy: Liguori.

Giddens, A. (1990). *The consequences of modernity*. Stanford, CA: Stanford University Press.

Godfray, H. C. J., Beddington, J. R., Crute, J. I., Haddad, L., Lawrence, D., Muir, J. F., et al. (2010). Food security: The challenge of feeding 9 billion people. *Science, 327*, 812–818.

Goodman, D., & DuPuis, E. M. (2002). Knowing food and growing food: Beyond the production–consumption debate in the sociology of agriculture. *Sociologia Ruralis, 42*(1), 5–22.

Gottschick, M. (2013). Reflexive capacity in local networks for sustainable development: integrating conflict and understanding into a multi-level perspective transition framework. *Journal of Environmental Policy and Planning*, 1–22.

Grasseni, C., Forno, F., & Signori, S. (2013). Beyond alternative food networks an agenda for comparative analysis of Italy's solidarity purchase groups (GAS) and districts of solidarity economy (DES) vis-à-vis US Community Economies. In *Proceedings of UNRISD Conference, Potential and Limits of Social and Solidarity Economy*. Geneva, Switzerland, 6–8 May 2013.

Guthman, J. (2007). The polanyan way? Voluntary food labels as neoliberal governance. *Antipode, 39*(3), 456–478.

Hazell, P., & Wood, S. (2008). Drivers of change in global agriculture. *Philosophical Transactions of the Royal Society B, 363*, 495–515.

IPCC. (2014). Climate change, impacts, adaptation, and vulnerability. Retrieved June 21, 2017 from http://www.ipcc.ch/report/ar5/wg2/.

Isin, E. F. (2009). Citizenship in flux: The figure of the activist citizen. *In Subjectivity, 29*, 367–388.

Keane, J. (2003). *Global civil society?*. Cambridge, UK: Cambridge University Press.

Lang, T., & Barling, D. (2012). Food security and food sustainability: Reformulating the debate. *The Geographical Journal, 178*(4), 313–326.

Leonardi, L. (2001). *La dimensione sociale della globalizzazione*. Rome, Italy: Carocci.

Lush, S. (1994). *Reflexive modernization*. Cambridge, UK: Polity Press, Cambridge.

Magatti, M. (2005). *Il potere istituente della società civile*. Rome, Italy: Laterza.

Maino F., Lodi Rizzini, C., & Bandera, L. (2016). *Povertà alimentare in Italia: le risposte del secondo welfare*. Bologna, Italy: Il Mulino.

Maslen, C., Raffle, A., Marriott, S., & Smith, N. (2013). *Food poverty: What does the evidence tell us?* Food Poverty Report, Bristol City Council, Bristol, UK.

Marsden, T. (2013). From post-productionism to reflexive governance: Contested transitions in securing more sustainable food futures. *Journal of Rural Studies, 29*, 123–134.

Marsden, T., & Sonnino, R. (2012). Human health and wellbeing and the sustainability of urban-regional food systems. *Current Opinion Environmental Sustainability, 4*(4), 427–430.

Monzini, P. (2007). Sea-border crossings: The organization of irregular migration to Italy. *Journal of Mediterranean politics, 12*(2), 163–184.

Moulaert, F., Martinelli, F., Swyngedouw E., & Gonzalez, S. (2005). Towards alternative model(s) of local innovation. *Urban Studies* (Vol. 42, No. 11, pp. 1969–1990), October 2005.

Oliveri, F. A. (2015). Network of resistances against a multiple crisis: SOS Rosarno and the experimentation of socio-economic alternative models. *Partecipazione e Conflitto, 8*(2), 504–529.

Pimbert, M., Tran-Thanh, K., Deléage, E., Reinert, M., Trehet, C., & Bennet, E. (2006). *Farmers' views on the future of food and small scale producers*. London, UK: International Institute for Environment and Development.

Pugliese, E., de Filippo, E., De Stefano, D., Dolente, F., Oliviero, L., & Pisacane, L. (2012). *Diritti Violati, Indagine sulle condizioni di vita dei lavoratori immigrati in aree rurali del Sud Italia e sulle violazioni dei loro diritti umani e sociali*. Napoli, Italy: Dedalus Cooperativa.

Pulcini, E. (2009). *La cura del mondo. Paura e responsabilità nell'età globale*. Torino, Italy: Bollati Boringhieri.

Renting, H., Marsden, T. K., & Banks, J. (2003). Understanding alternative food networks: Exploring the role of short food supply chains in rural development. *Environment and Planning, 35*, 393–411.

Retegas. (1999). Documento base dei GAS. I Gruppi di Acquisto Solidali. Un modo diverso di fare la spesa. Retrieved June 21, 2017 from www.retegas.org/upload/dl/doc/GASDocumentoBase.pdf.

Savioli, A., Crisci, G., & Fonte, M. (2015). Il Papavero: The butterfly laboratory and the financial strength of solidarity economy. In *Proceedings of Second International Conference on Agriculture in an Urbanizing Society: Reconnecting Agriculture and Food Chains to Societal Needs, 14–15 September 2015*. Rome, Italy.

Schifani, G., & Migliore, G. (2011). Solidarity purchase groups and the new critical and ethical consumer trends: First result of a direct study in Sicily. *New Medit, 11*(3), 26–33.

Sen, A. K. (1981). *Poverty and famines: An essay on entitlement and deprivation*. Oxford, UK: Clarendon Press.

Sen, A. K. (1995). *Inequality reexamined*. Harvard, USA: Harvard University Press.

Sen, A. K. (1999). *Development as freedom*. New York, USA: Oxford University Press.

Sen, A. K. (2000a). *Freedom, rationality, and social choice: The arrow lectures and other essays*. Oxford, UK: Oxford University Press.

Sen, A. K. (2000b). Food entitlements and agricultural production. In *Proceedings of Doctoral Lecture, 30 November 2000*. University of Florence, Italy.

Sen, A. K. (2012). The snakes and ladders of europe. *The economic system we need*, 10 May 2012.

Shetty, P. (2015). From food security to food and nutrition security: Role of agriculture and farming systems for nutrition. *Current Science, 109*(3), 456–461.

Sonnino, R., Moragues Faus, A., & Maggio, A. (2014). Sustainable food security: An emerging research and policy agenda. *International Journal of the Sociology of Agriculture and Food, 21* (1), 173–188.

Sonnino, R. (2016). The new geography of food security: Exploring the potential of urban food strategies. *The Geographical Journal, 182*(2), 190–200.

Sonnino, R., & Hanmer, O. (2016). Beyond food provision: Understanding community growing in the context of food poverty. *Geoforum, 74*, 213–221.

Stirling, A. (2009). Direction, distribution and diversity! pluralising progress in innovation, sustainability and development. In *STEPS Working Paper 32*. Brighton, UK: STEPS Centre, University of Sussex.

United Nations. (1974). Universal Declaration on the Eradication of Hunger and Malnutrition. In *World Food Conference General Assembly, 5–16 November 1974*. Rome, Italy.

United Nations. (2015). Transforming our World: 2030 Agenda for Sustainable Development. In *70th Session of the General Assembly United Nations, 25–27 September 2015*, United Nations Headquarters. New York, USA.

Voss, J. -P., Kemp, R. (2005). Reflexive governance for sustainable development—Incorporating feedback in social problem solving. In *Proceedings of IHDP Open Meeting*, 9–13 October 2005. Bonn, Germany.

Voss, J. P., & Kemp, R. (2006). *Sustainability and reflexive governance*. Cheltenham, UK: Edward Elgar.

Voss, J. -P., & Bornemann, B. (2011). The politics of reflexive governance: Challenges for designing adaptive management and transition management. *Ecology and Society, 16*(2), Art. 9.

Wolff, F. (2006). The transformation of agriculture: Reflexive governance for agrobiodiversity. In J. -P. Voss, D. Baukecht, & R. Kemp (Eds.), *Reflexive governance for sustainable development* (pp. 383–416). Cheltenham, UK: Edward Elgar Eds.

Daniela Bernaschi is a development economist and Ph.D candidate in Political and Social Change at the University of Turin and Florence, Italy. She deals with food insecurity in Europe and the role played by civil society organizations, using the Capability Approach by Amartya Sen. She is currently visiting student at the Autonomous University of Barcelona, Spain.

Giacomo Crisci is a development economist and MSc candidate in Food, Space and Society at Cardiff University, United Kingdom. He is lecturer in the Master of Arts in Food Studies at the American University of Rome. His work is centred around the solidarity purchase groups in Italy and the challenge of scaling up social and solidarity economy.

We would like to thank Dr. George Baourakis and Prof. Konstadinos Mattas for organizing the 158th EAAE Seminar on Food Security and Sustainable Agriculture for which this article has been written.

We are particularly grateful to our colleague Ms. Benedetta Tavani, who commented on an early draft of this article. Special thanks also to Gabriella, Luigi, Elisabetta and Francesca for their inestimable support.

The usual disclaimer applies.

Role of Buffalo Production in Sustainable Development of Rural Regions

Ibrahim Soliman

Abstract Rice is the main summer crop in Egypt. It is a cash exportable crop that provides a main source of income to the Egyptian farmers and the national economy. However, the farmers used to burn the rice straw at the farm borders and violate the law that forbids such action, which causes socio-economic negative externalities due to the generated smoke from burning. The smoke generated from burning is straw produced as byproduct of cultivated around 0.75 million ha of rice crop in Egypt, causes social costs due to the probability of premature-mortality and morbidity of rural and urban individuals and livestock. To conduct an economic assessment of such negative externalities a field research was conducted. A targeted ration of chopped rice-straw mixed with dissolved urea and molasses at 2% and 3% of weight, respectively, was fed to buffalo-feeder calves for meat production at 40% of the S.E. of the daily ration with a concentrate feed mix of 60% S.E. Such ration was compared with a control ration of dray chopped rice straw with the same proportion of concentrate feed mix. Two feed-response models were estimated for comparison of the two rations on the growth of the buffalo feeder calves for meat production. The Cobb-Douglas response function was the best fitted form according to the economic logic, significance of estimated parameters and the magnitude of R-2. The study derived the production elasticity, marginal daily gain, the value of marginal product from both estimated feed response functions. The economic marketing weight that maximizes the gross margin above the feed costs was estimated under the response model of treat rice straw feeding plan (targeted ration). It reached around 518 Kg live weight, while under the control ration it was only around 384 Kg. The larger market weight of treated rice straw ration was due to higher production elasticity, faster marginal daily gain, better marginal feed conversion and higher palatability of the ration than the control one. Egypt imports of red meat reached about 600 million dollars, due to lack of sufficient feed supply that constrained expansion in red meat production. Therefore, providing treated rice straw feed would provide additional source of livestock feeds which would provide additional 80,000 tons' carcass weight from fed buffalo calves, which currently are

I. Soliman (✉)
Agricultural Economics, Faculty of Agriculture, Zagazig University, Zagazig, Egypt
e-mail: ibrahimsoliman12@gmail.com

© Springer International Publishing AG, part of Springer Nature 2018
K. Mattas et al. (eds.), *Sustainable Agriculture and Food Security*,
Cooperative Management, https://doi.org/10.1007/978-3-319-77122-9_2

slaughtered as rearing veal calves (60-80 days old). The estimated income generated from one buffalo fed calves reached 50% of the average annual per capita income in Egypt. Such program would also stop the social costs stems from probable premature death and/or morbidity of human and livestock when burning rice straw. The study presented a proposed institutional program to introduce such technology into Egyptian agricultural sector.

Keywords Externalities of burning rice Straw · Recycling of treated rice straw as feed · Buffalo-Feed response function · Least cost ration of alternative feeding systems · Most Profitable marketing weight

This chapter focused on the socio-economic evaluation of using treated-rice straw as fodder for fattening buffalo male calves for three feasible reasons: (1) The Egyptian budget has faced an increasing burden due to the speedy increase of net imports of red meat, while the value of the Egyptian pound is decreasing fast in front of the dollar, at least over the last four years, (2) more than 70% of buffalo calves are slaughtered at the rearing period to save fodders for dairy buffaloes. (3) to use treated rice straw as feed would protect the society from probable harmful impacts of burning abundant rice straw on the environment and human health. The socio-economic impacts were derived from two estimated econometric models for buffalo-feed response function. The data were collected from designed field experiments.

Therefore, this chapter dealt with the following issues: Impacts of the economic policies in Egypt on farmers' usage of rice straw; Negative effects of open-field rice straw burning; Alternative recycled products of rice straw; Importance of buffalo in protein production systems in Egypt; Role of treated rice straw in bovine feeding systems; Identification of the research problem; Major objective of the presented research work; Sources of data base; The implemented econometric model's structure and derived function forms; The results were presented under response model of the control feeding system and the response model of the alternative system. The discussion of the results included: (1) The maximum gross margin and optimum marketing weight, (2) the techno-economic efficiency of the feed response under the two feeding systems, (3) socio-economic impacts of the treated rice straw feeding system for buffalo fattening, (4) the biological and economical interpretations of the higher efficiency of rice straw silage than chopped dry straw. The chapter was ended by major conclusions and recommendations for the concerned institutions in the agricultural sector.

1   Impacts of the Economic Policies in Egypt on Farmers' Usage of Rice Straw

Till the early eighties of the last century, wheat straw was the major roughage feed for Egyptian livestock and farmers had negative views towards using rice straw for livestock feeding. Till that time, Egypt was practically under a planned economy system. Whereas the domestic and imported wheat price and supply distribution were controlled by the government, wheat straw was at free market price. Therefore, up to the seventies of the last century, its price surpassed wheat grain price for many years (Soliman and Nawar 1986). This was mainly due to the limited area cultivated with wheat and the low yield of wheat grains, leading to a low yield of wheat straw. The wheat area in Egypt was 1,395,382 feddans in 1961, (1 feddan = 4200 M^2), due to controlled cropping patterns, and the yield per feddan, at that time, was about 1.029 tons (FAOSTAT 2015). Since 1986/1987, Egypt has started an economic reform program which implied liberalization of the prices and marketing of crops and agricultural inputs (Hezall et al. 1995). The program was associated with introducing new high yield varieties of grains and expansion in agricultural mechanization systems. Therefore, the yield per feddan of wheat increased rapidly to reach about 2.778 tons per feddan in 2013, associated with expansion in the wheat area to reach 3,404,899 feddans due to the free decision to cultivate and provide high farm guaranteed price. Since then, wheat straw production and supply has increased to a great extent. The major summer crop, i.e. rice, was exposed to the same policies and then showed the same performance over the same period. Its area and yield increased from 941,667 feddans and 1.213 tons per feddan, respectively, in 1961 to 2,916,667 feddans and 2.314 tons per feddan, respectively, in 2013. Accordingly, the farmers have been faced recently with an abundant supply of rice straw, as the farmers still prefer to use wheat straw rather than rice straw as feed. Thus, the majority of them preferred to burn the rice straw at the farm border, which caused the phenomenon of "the black cloud" all over the Nile delta governorates during the rice harvesting season (September–October) of every year. As burning of rice straw has harmful impacts on the environment and human health, farmers have thus been encouraged to refrain from burning rice straw and to adopt more environmentally safe and human-friendly rice straw management practices.

2   The Negative Effects of Open-Field Rice Straw Burning

Burning causes atmospheric pollution and results in nutrient loss, even though it is a cost-effective method of straw disposal and helps reduce pest and disease populations (Dobermann and Fairhurst 2002) The environmental consequences of rice straw burning in terms of greenhouse gas (GHG) emissions are, mainly, carbon dioxide, methane and nitrous oxide gases, that require adoption of selected rice straw management alternatives (Launio et al. 2013). The compositions of biomass are variable, especially with respect to inorganic constituents. Alkali and alkaline earth metals, in combination with silica, sulfur and chlorine, are responsible for many undesirable reactions in the combustion of straw (Jenkins 1999).

Burning of rice straw causes almost complete N loss, P losses of about 25%, K losses of 20%, and S losses of 5–60%. The proportion of nutrients lost depends on the method used to burn the straw. When straw is heaped into piles at threshing sites and burned after harvest, the ash is usually not spread on the field, resulting in large losses of minerals K, Si, calcium (Ca) and magnesium (Mg) leached from the ash piles.

3 Alternative Recycled Products of Rice Straw

There are several approaches to utilizing rice straw nutrient components. The most common approach is to be removed from the field, burned in situ, piled or spread in the field, incorporated in the soil, or used as mulch for the following crop. About 40% of the nitrogen (N), 30–35% of the phosphorus (P), 80–85% of the potassium (K), and 40–50% of the sulfur (S) taken up by rice remains in vegetative plant parts at crop maturity. Each of these measures has a different effect on overall nutrient balance and long-term soil fertility. When straw is the only organic material available in significant quantities to most rice farmers and where S-free mineral fertilizers are used, straw may be an important source of S; thus, straw burning should not be practiced. However, spreading and incorporation of straw are labor-intensive tasks and farmers consider burning to be more expedient (Dobermann and Fairhurst 2002).

In contrast, burning effectively transforms straw into a mineral K nutrient source, and only a relatively small amount of K is lost in the process. Therefore, the effect of straw removal on long-term soil fertility is much greater for K than for P (Launio et al. 2013). Straw is also an important source of micronutrients such as zinc (Zn) and has the most important influence on the cumulative silicon (Si) balance in the soil (Nelson et al. 1980).

In addition to being a fertilizer, rice straw can be used as fuel for cooking, ruminant fodder, and stable bedding or as a raw material in industrial processes, e.g., paper making (Nelson et al. 1980).

4 Importance of Buffalo in Protein Production Systems in Egypt

The Egyptian buffalo is a river buffalo type which is a milk and meat producing animal. The number of buffalo stock increased from 2.897 million heads in 1990 to more than 3.949 million heads in 2014, (FAOSTAT 2016), i.e. at an average annual growth rate of 2.2%[1]. The buffalo meat production in Egypt was about 161 thousand tons of carcass weight in 1990, i.e. about 41.7% of the total red meat production in 1990. Even though buffalo meat production increased to about 390 thousand tons of carcass weight in 2014, its share in total red meat production had not changed much, i.e. it amounted to 41% due to slaughtering most of the male calves at rearing age. This behavior stems mainly from the shortage in feed availability and to save buffalo milk yield that receives a high sale price because of consumer preferences, (Soliman 2006).

---

[1]Calculated from the exponential function: $y_t = y_0 e^{rt}$ where y_t = value of the concerned variable in the year t, y_0 = the value in the base year, it is the digital number of years and r is the average annual growth rate which is estimated from: $r = [Ln(y_t) - Ln(y_0)]/t$

With respect to buffalo milk supply in Egypt, the production of milk was about 1.250 million tons in 1990 and this increased to more than 2.500 million tons in 2014 (FAOSTAT 2016) i.e. an annual growth rate of about 4.5%[1]. It represents about one-half of total milk production in Egypt. However, due to its higher total solids, particularly fat (8%), readjusting the volume of buffalo milk production on the basis of equivalent milk with 4% fat, its share in total milk production would reach more than two-thirds. Due to the high total solids of buffalo milk, it is a main source of domestic "feta" cheese.

The cost of milk production from buffaloes is also less than the cost of reconstituted imported powdered milk at the international market price and the cost of milk from domestic cattle. It is the most important and popular livestock for milk production in Egypt. Buffalo productivity in Egypt is about 210–280 days/lactation, an average of seven lactations and milk yielding 1,600 kg/lactation season with 7–8% fat. The age at the first calving is 32–38 months. The average annual red meat production contributes 45% to the overall meat produced. Artificial insemination is used in one percent of the medium to large herds. There are six artificial insemination (AI) stations owned by the Government and one by the University, possessing a total of 70 bulls. AI is still performed at research level; usually only one semen dose is offered at each estrus, conception at the first estrus being 30%. Milking is done by hand, twice a day, mainly by women. Average slaughter weight is 500 kg, at the age of 18–24 months. Carcass yield is 51%. Overall growth rate is 700 g/day, (Ibrahim 2012). Buffalo exceed cattle in their ability to convert poor quality and forage to meat or milk, (Soliman and Nawar 1986).

5 Role of Treated Rice Straw in Bovine Feeding Systems

Feed resources and nutrition constitute the principal technical constraints to ruminant production in Asia. Among recommended non-conventional feed resources is the option of improving the nutritive value of crop residues. More attention has been given to the chemical treatment of cereal straws than to supplementation. However, failure to demonstrate cost-effectiveness has discouraged on-farm adoption, even though there is significant potential for the more effective use of locally-produced non-conventional feed resources (Devendra and Sevilla 2002).

Tengyun (2000) reported that China produces more than 500 million tons of crop straw and Stover every year. By promoting feeding of livestock with crop straw and Stover, the beef and mutton output could be increased markedly and a great amount of feed grain would be saved. The researcher reviewed the different techniques for the utilization of crop straw and Stover. The techniques that have been used in China include physical treatment (chopping), chemical treatment and microbial treatment methods.

Ruminant production plays an important part in the predominantly agricultural economy of Vietnam, especially in mixed animal-crop production systems. These animals themselves largely rely on crop residues as feed. The scarcity of land and the trend of sustainable agricultural development in the highly-populated country necessitate better utilization of crop residues in general and particularly, rice straw for ruminant feeding. Although rice straw produced every year is plentiful, the amount a ruminant can consume is not sufficient to sustain a reasonable level of production due to its low nutritive value. Therefore, rice straw has not been maximally utilized for ruminant production yet. Suitable treatment techniques in combination with nutrient supplementation could result in improved utilization of rice straw with better benefits. Despite recent local research in this field, no methods for improved utilization of straw are practically applied by farmers in the country, probably because none has proved to be relevant and sustainable under the local physical and socio-economic conditions (Trach 1998).

After conducting a preliminary survey, Duc Vu et al. (1999) carried out a feeding trial in Vietnam to determine the effect of urea–molasses–multi-nutrient block (UMMB) and urea-treated rice straw (UTRS) as a feed supplement on the productivity of dairy cows. Sixty Holstein–Friesian crossbred cows on 11 small-holder farms were divided equally into control, UMMB and UTRS supplementation groups. Milk yield and feed intake were recorded daily. Milk fat content and the body weight of each cow were determined at two week intervals. Recorded data showed that milk production increased by 10.3–11.9% and milk fat content increased by 3–5%; therefore, profit for farmers increased by US $0.55–0.73 per cow per day.

Akter et al. (2015) conducted an experiment to study the effect of treatment of rice straw with urea and a urease containing midden soil on feed intake of the animals, nutrient digestibility, body weight gain, feed conversion efficiency and the overall economy of feeding for a period of 105 days. Twelve indigenous growing cattle (live weight 130.00 ± 1.67 kg) were selected and divided into four groups having three animals in each group. The animals received 3.0% urea + 2.0% treated rice straw (group A), 3.0% urea + 3.0% midden soil treated rice straw (group B), 3.0% urea + 4.0% midden soil treated rice straw (group C) and 3.0% urea + 5.0% midden soil treated rice straw (group D). In addition, all the animals were supplied with 2 kg green grass, 450 g concentrate mixture and 40 g salt per 100 kg body weight. The total live weight gain by the end of the experimental period (105 days) was 39.00, 42.50, 46.50 and 49.00 kg for groups A, B, C and D respectively. The addition of 5.0% midden soil as a urease source with 3.0% urea (D) treated rice straw significantly ($P < 0.01$) increased the coefficient of digestibility in all nutrient intakes. The total profit of meat production in group D was significantly higher ($P < 0.01$) than in groups A, B and C.

Chemjong (1991) studied the effects of feeding urea-treated rice straw to lactating buffaloes in Nepal. Six pairs of similar buffaloes on farms were selected. All of them were given a conventional diet based on rice straw for four weeks, then one of each pair was given 15–20 kg/day of urea-treated rice straw for a period of 4 weeks while the control group received untreated rice straw. In the final 4 week period, all animals were given a conventional diet. Feeding straw treated with 4% urea increased the voluntary intake of straw by 25% and increased milk yield by 1.6 L/day compared with buffaloes fed the conventional diet containing untreated straw. Milk production remained elevated after the four-week treatment period had finished. The results showed that buffalo cows fed urea-treated straw achieved better weight gain, and milk yield increased significantly (P < 0.01) compared with the control animals. During the treatment period the net benefit was US$ 0.16 per day and the incremental rate of return was 46%. Moreover, in the 4 weeks following the treatment period the net benefit was US$ 0.40 per day. Ensiling rice straw with 4% urea, i.e. making silage, can be recommended as a safe, economical and suitable method for improving the nutritional value of rice straw on small farms in Nepal, thus increasing milk production and the live weight of lactating buffaloes. The practice of feeding urea-treated straw is economical for farmers during the dry season from January to April.

Naik et al. (2004) conducted a study on the effect of feeding ammoniated wheat straw treated with HCL and without HCL on meat quality and various sensory attributes of growing male buffalo calves. Due to urea-ammonization, the Crude Protein (CP) content of wheat straw increased from 2.90 to 6.96%. The addition of HCL along with urea further increased the CP content to 10.09%. The proximate composition as a percentage of the fresh first quality meat cuts (muscles) was comparable among the groups. Results indicated the desirable effect of feeding either without or with HCL. The scores of the cooked (2% common salt) for various sensory attributes (appearance, flavor, juiciness, texture, mouth coating and overall palatability) were comparable among the groups. In conclusion the results suggested that feeding of ammoniated wheat straw treated with and without HCL to growing male buffalo calves for 180 days had no adverse effect on the meat quality and various sensory attributes.

Rahman et al. (2009) carried out an experiment to compare a ration combination of urea-molasses-straw with green fodder versus other ration combinations, where they replaced the basic ration ingredients with different levels of concentrate feed. They found a significant tendency of increase in live weight gain, better nutrient digestibility, and improvement in feed conversion rate with increased levels of concentrate supplementation.

Fifteen calves of Baladi cattle and a similar number of buffalo calves in Egypt were used to compare daily gain, feed conversion and costs of producing meat from both types (El Asheeri 2012). Calves were purchased from the local market with a body weight of 229–238 kg. Calves were kept tied in a semi-open yard and fed on concentrate feed mixture and rice straw throughout the fattening period. The initial body weight of Baladi and buffalo calves was similar, averaging 231.4 ± 1.9 and 232.6 ± 1.0 kg, respectively. Calves were kept growing up to the final body weight of 400 kg and the monthly body weight was recorded. Growth curve, average daily gain and total weight gain were also recorded during the fattening period. The cost of producing one kg gain and one kg meat were calculated as economic parameters. The growth curve of the two studied genotypes indicated no significant difference between Baladi and buffalo calves within the first three months of the fattening period. Afterwards, the body weight of Baladi calves increased significantly compared to buffaloes. The fattening period of buffalo calves (252.7 ± 5.7 day) was significantly longer than that of Baladi (185.7 ± 7.4 day) by about 67 day.

This was due to the higher average daily gain of Baladi (0.93 kg) than buffaloes (0.67 kg) by about 38.8%. Total operational costs of fattening buffalo calves were significantly higher than those of Baladi by about 35.9%. The feed conversion rate of buffalo calves (13.1 kg dry matter) was 37.9%, significantly higher than for Baladi calves (9.5 kg dry matter) to reach the target body weight of 400 kg. The cost of producing one kg weight gain in buffalo calves was LE 15.4, i.e. higher than that of the cost of Baladi calves of LE 11.2. The corresponding cost of producing one kg meat was LE 23.8 and 35.9 for Baladi and buffalo calves, respectively.

To investigate the effect of concentrate levels of the daily ration combination on the performance of buffalo calves in Egypt, Helal et al. (2011) carried out an experiment on twenty-one male buffalo calves. The average initial body weight was about 286 kg. They were divided into three equal groups and randomly assigned to the following concentrate levels: 70, 85 and 100% of the concentrate feed mixture. They were referred to as groups A, B and C, respectively. Allowance of the concentrates was offered daily in equal portions while roughage (rice straw) was available at all times. Body weight, feed intake and feed conversion were determined. At the end of the experiment all calves were slaughtered and carcass traits were recorded. No differences between groups regarding the change in body weight were detected. Calves fed 100% concentrate gained more than the other two groups in all experimental periods. The overall average feed intake for groups A, B and C was 8.65, 9.81 and 11.11 kg/d, respectively. Feed conversion was better for the 70% group than the control. Heavier weights of carcass, bone and boneless meat were obtained from calves of group C compared to groups A and B. However, such differences were insignificant. Dressing percentage was significantly higher in groups B and C than group A. the meat-to-fat ratio followed the same pattern of dressing percentage. It seems that the close concentrate proportion between the three groups was behind such results of no specific trend in differences.

6   Identification of the Research Problem

This chapter focused on the socio-economic evaluation of using treated-rice straw as fodder for fattening buffalo male calves for three feasible reasons. First, the Egyptian budget has faced an increasing burden due to the speedy increase of net imports of red meat, from around 303 million dollars in 2000 to more than 963 million dollars in 2011 (FAOSTAT 2015), while the value of the Egyptian pound is decreasing fast in front of the dollar, at least over the last 4 years.

The exchange rate was less than 5.6 EGP/ $1 in 2009 and reached 7.47 EGP/$1 in January 2015 (Central Bank of Egypt 2015). Secondly, the buffalo population in Egypt yields around 700,000 heads of male calves. At least 500,000 are slaughtered at the rearing period (around 2–3 months old) to save both buffalo milk for sale and green fodder (clover) in winter for dairy buffalo feeding. Such numbers of buffalo veal yield around 20,000 tons of carcass weight (40 kg per head). If such a large number of buffalo calves were kept for feeding, they yield around 100,000 tons of carcass weight (about 200 kg carcass weight/head at around 24–30 months old). The net added meat production would be 80,000 tons of carcass weight. Such an amount might share significantly in filling in the market gap between domestic production and demand for red meat in Egypt.

7 Major Objective of the Presented Research

Therefore, the empirical objective was to estimate an econometric model for the feed response using a ration composed of treated rice straw silage with the common concentrate feed mix for feed-lot system of male buffalo feeder calves to reach the optimum marketing weight which maximizes the gross margin above the least cost ration.

7.1 Data Base

The author used the data of a field sample survey from progressive large livestock farms in the north-east Nile delta region of Egypt. The sample was composed of 60 buffalo male calves. A subsample of 30 heads were fed chopped rice straw with concentrate feed mix as a control (Feeding System 1) and the other 30 heads were fed chopped rice straw silage treated with dissolved urea (2% of the straw weight) and molasses (3% of the straw weight) with concentrate feed mix (Feeding System 2). The data covered the agricultural year 2013–2014. The ration combination in both systems as Starch Equivalent quantity (kg S.E.) was 60% concentrate feed mix and 40% rice straw. The cumulative live weight till marketing with the associated feed intakes data was recorded bi-weekly. The feeder calf weight (initial weight) was around 185 kg live weight. The annual average input and output prices were collected for the year (2013/2014), as shown in. Table 1, the concentrate feed mix was composed of corn, oilseed meal, bran and molasses. Its S.E. equivalent weight is 0.52%, and 12% crude protein of the natural form weight.

Table 1 Average price of inputs and outputs for buffalo meat fattening (2013/2014)

Item	Fed buffalo male calves (400 kg/Head)	Concentrate feed mix	Rice straw	Rent of chopping machine	Urea	Molasses
Unit	1-Kg Live weight	1-Ton	1-Load = 250 kg	1 h for 1 ton straw	11-Sac = 50 kg	Tin (30 kg)
Egyptian pound (EGP)	28.75	2800	100	60	100	10

1-$ = EGP 7.47

7.2   Methods and Analytical Procedures

A feed response model was estimated for the two feeding systems. However, the two feed items were fed at a constant combination, i.e. 40% rice straw and 60% concentrate feed mix as S.E. equivalent value. Therefore, the estimated linear correlation between the intake quantities of both feed items was about 0.924. It was evidence of a multicollinearity problem (Intriligator 1978), which might cause biased estimates of the production response model and violation of the statistical inferences if both feed variables were introduced to such a model as explanatory variables. Accordingly, the study aggregated the feed items as one explanatory variable that expressed feeding level as S.E. in kg.

The literature showed that the livestock feed response is often not a linear relationship. It follows the principal of dimensioning return of inputs (Soliman 2006). Therefore, the best fitted feed response function was the quadratic function, as shown by Eq. (1):

$$\acute{y}_i = \alpha + b_1 x - b_2 x^2 \tag{1}$$

Several functions were derived from the response function (1) to estimate the technical and economic efficiency of each feeding system. Equation 2 is the estimated physical marginal product "MPP" The average physical production (APP) is presented by Eq. 3. The production elasticity function (Eq. 4) is derived from Eqs. 2 and 3. The estimated optimum marketing weight that recognizes the maximum gross margin above the least cost feed intake (Eq. 5) is derived from Eq. 2 to express the condition of the equilibrium economic point when the value of marginal product (VMP) equals the marginal cost (MC) i.e., the price of 1 kg S.E of the ration. The market prices are presented in Table 1 and the economic efficiency coefficient "EE" (Eq. 6) is derived from Eq. 5.

In addition, the model of the fattening time function as presented by Eq. 7 was used to estimate the required period to reach the optimum marketing weight. The first derivative of Eq. 7 is the marginal time required to consume an additional unit of feed (Eq. 8). From the two parametric Eqs. (2 and 8) a Cartesian equation (Eq. 9) is derived which estimates the marginal daily gain of live weight. The gross margin (Eq. 10) presents a measure of profitability per head of one fed buffalo calf. It is not the normal profit, but the margin left above the feed cost:

$$MPP_x = dy/dx = b_1 - 2b_2 x \tag{2}$$

$$APP_x = Y/X \tag{3}$$

$$\xi_x = MPP_x/APP_x \tag{4}$$

$$\text{Maximum gross margin when: } VMP_x = P_y(MPP) = P_x \qquad (5)$$

$$EE = VMP_x/P_x \qquad (6)$$

$$T = c_0 + C_1 X + C_2 X^2 \qquad (7)$$

$$MPP_{x \cdot t} = dT/dx \qquad (8)$$

$$MPP_t = MPP_x/MPP_{x \cdot t} = dy/dx \quad dT/dx \quad \frac{dy}{dx} dx/dy \quad dy/dx. \qquad (9)$$

$$GM = P_y(MW) - \sum P_{xi}X_i \qquad (10)$$

where:

$\acute{y}$ Estimated cumulative weight gain in kgs live weight of calf I

α Estimated intercept

x Feed intake in kg of S.E. (40% rice straw and 60% Conc. mix)

b_i estimated feed response coefficient (regression coefficient)

MPP_x marginal physical product estimates the additional weight gain due to additional 1 kg of ration

APP_x estimates the average live weight gain per 1 kg of ration

ξ_x Production elasticity = the relative change in weight gain due to 1% change in feed intake

VMP_x estimates the marginal revenue per additional 1 kg of feed combination

P_y Price of 1 kg live weight in EGP

P_x Price of 1 kg S.E. of ration in EGP

EE Economic Efficiency coefficient = marginal revenue generated by spending an additional 1 EGP on feeds. If it is >1, this means it is feasible to expand in feed use for more weight gain; if it is less than one, then less feed should be used and less marketing weight is more feasible.

T time period of feeding a calf till the marketing weight in days

C_i estimated regression coefficients of the time function of the feed consumption

$MPP_{x \cdot t}$ marginal time for each additional feed unit intake

MPP_t Marginal Daily Gain

GM Gross Margin in EGP = Total revenue − Feed costs

7.3   *Results and Discussion*

The estimated response functions and the derived functions are presented under each feeding system, with associated statistical inferences estimates.

The values under estimated parameters are the SE. of these estimates.

8 Feeding System 1: Chopped Rice Straw with Conc. Feed Mix

$$\acute{y} = 120.155 \quad +0.283x \quad -0.000089x^2 \tag{11}$$
$$\quad\quad\quad (6.16) \quad\quad\quad (-2.29) \quad\quad R^{-2} = 0.93 \quad F = 185.05$$

$$MPP_t = (0.283 - 0.00018X)/(0.405 - 0.000024X) \tag{12}$$

$$MPP_x = 0.283 - 0.00018X \tag{13}$$

$$T'_1 = 4.972 \quad +0.405X \quad -0.000012X^2 \tag{14}$$
$$\quad\quad (80.23) \quad\quad\quad (-17.29) \quad\quad R^{-2} = 0.988 \quad F = 2372.1$$

9 Feeding System 2: Rice Straw Silage with Urea and Molasses and Conc. Feed Mix

$$\acute{y} = 0.902 + 0.349x \quad -0.000079x^2 \tag{15}$$
$$\quad\quad\quad (17.14) \quad (-2.98) \quad\quad R^{-2} = 0.985 \quad F = 1634.1$$

$$MPP_x = 0.349 - 0.00016X \tag{16}$$

$$T'_1 = 4.075 \quad +0.386X \quad -0.00012X^2 \tag{17}$$
$$\quad\quad (99.95) \quad\quad\quad (-23.99) \quad\quad R^{-2} = 0.988 \quad F = 2372.7$$

$$MPP_t = (0.349 - 0.00016X)/(0.386 - 0.00024X) \tag{18}$$

Table 2 presents the estimated techno-economic criteria of fed calves under the two tested feeding systems at the price levels of inputs and outputs shown in Table 1. The results are presented in two sections: the first compares the optimum live weight, least cost ration and gross margin of the two systems, and the second section provides a comparative analysis of the estimated average techno-economic criteria of the two feeding systems.

10 Maximum Gross Margin and Optimum Marketing Weight

At the market price level of feeds and live weight of buffalo fed calves in 2013/2014, the estimated marketing weight that maximizes the gross margin above the feed costs was 383 kg live weight under Feeding System 1, while it reached about 517 kg live weight under Feeding System 2 (Table 2 and Fig. 1). Both estimated

Table 2 Economic analysis profile of the fed buffalo calves on two different feeding systems

Estimated techno-economic criteria	Unit	Feeding system 1	Feeding system 2
Average marginal live weight gain/ Kg S.E	Kg live weight	0.150	0.231
Production elasticity of feed intake	%Gain/1% feed	0.72	0.84
Least cost ration quantity	Kg S.E.	872	1381
Optimum marketing weight	Kg live weight	383	517
Feeding period to reach marketing weight	Days	436	382
Total revenue	EGP	10632	14354
Feed costs	EGP	3092	5012
Gross margin	EGP	7540	9342
Average marginal daily gain/day	Kg live weight	0.351	0.881
Average daily gain	Kg live weight	0.21	0.27
Average economic efficiency	EGP	1.17	1.76

Source Estimated from: Eqs. (10–17), Table (1) and application of the Cartesian Eqs. (5, 6 and 9)

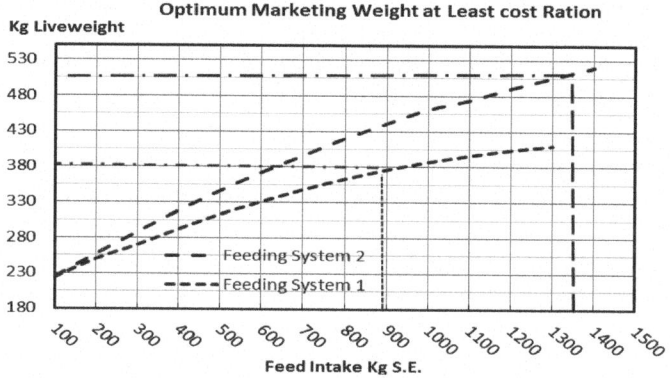

Fig. 1 Optimum marketing weight at least cost ration

weights were achieved at the least cost ration which maximizes the gross margin. Obviously, Feeding System 2 reaches a higher marketing weight at a larger quantity of kg S.E. than Feeding System 1 of less marketing weight, i.e. 1381 kg S.E. versus 872 kg S.E, respectively. Therefore, the least cost ration value of Feeding System 2 was higher than that of Feeding System 1, i.e. 5012 EGP and 3092 EGP, respectively. The higher price per kg S.E. of Feeding System 2 compared to Feeding System 1, i.e. 3.64 EGP versus 3.54 EGP, respectively was also behind the higher feeding cost under Feeding System 2. However, the higher total revenue at the larger marketing weight of Feeding System 2 much surpassed that of Feeding

System 1, i.e. 14353 EGP and 10632 EGP, respectively. Therefore, the acquired gross margin under Feeding System 2 reached 9342 EGP while that generated under Feeding System 1 was 7540 EGP. In other words the farmer would acquire a gross margin under Feeding System 2 of about 124% of what he/she could reach under Feeding System 1. As both subsamples had started with feeder calves of the same initial weight (185 kg/head) at an average age of calves around 14 months old, therefore, such higher marketing weight and gross margin generated by Feeding System 2 were due to better techno-economic performance resulting from the nutrient content of the second system, as discussed in the following section.

11   Techno-Economic Efficiency of the Feed Response Under Two Feeding Systems

The estimated production elasticity showed that while a 10% increase in feed intake under Feeding System 1 raises the live weight gain by 7.2%, it raises the live weight gain under Feeding System 2 by 8.4%. On average, an additional 1 kg S.E. of the ration under Feeding System 1 adds only 0.15 kg live weight gain, but adds 0,231 kg live weight gain under Feeding System 2. The buffalo fed calf would reach an optimum marketing live weight of 383 kg in 436 days under Feeding System 1 and an optimum marketing weight of 517 kg live weight in 517 days under Feeding System 2. Accordingly, the estimated average daily gain was 0.21 kg under Feeding System 1 and 0.27 kg under Feeding System 2. However, the average marginal daily gain from the time function is a better measure for growth speed. It is the average increase in live weight per additional day of growth. While such gain reached 0.351 kg under Feeding System 1, it was 0.881 kg under Feeding System 2, i.e. under Feeding System 2, the buffalo calves grow faster.

As the technical criteria of Feeding System 2 highly surpass those criteria under Feeding System 1, the estimated average economic efficiency coefficient under Feeding System 2 was significantly higher than such a coefficient under Feeding System 1, i.e. 1.76 and 1.17, respectively. It means that while each additional 1 EGP of ration costs under Feeding System 2 generates marginal revenue of 1.76 EGP, it generates only 1.17 EGP under Feeding System 1.

12   Socio Economic Impacts of the Feeding System (2) for Buffalo Fattening

Each buffalo calf fed a ration combination of concentrate feed mix with rice straw silage enriched with dissolved urea and molasses generates direct and external benefits to the rural communities: (1) it generates additional annual income to the farm household of about $1251, equivalent to one-half of the average per capita income in

Egypt, i.e. $2500 (World Bank 2013); (2) it generates extra employment opportunities for the family labor on the farm, which currently suffers from high unemployment (Soliman 2004); (3) it shares in decreasing the annual red meat imports; (4) it utilizes potential livestock resources by stopping the slaughtering of buffalo veal calves during the rearing period; (5) it stops the burning of rice straw which enables communities to avoid the probability of premature mortality and morbidity of not only rural but also urban human resources; and (6) some studies have measured the social costs of the probability of losing one's life by $2,00,000, (Soliman 1995).

13 Causes of the Higher Efficiency of Rice Straw Silage Than Chopped Dry Straw

It seems that the high techno-economic efficiency of the ration composed of 60% starch equivalent from concentrate feed mix and 40% from chopped rice straw silage (with 2% dissolved urea and 3% molasses) is due to some physiological reasons: (1) the enriched rice straw with dissolved urea raises the protein content of the ration; (2) addition of molasses activates the bacterial activities, during silage preparation, which raises the digestibility of the rice straw, and therefore its starch equivalent value; and (3) making silage from rice straw increases its palatability and gives it a preferable smell to the animals. Therefore it increases the intake of the rice straw silage.

13.1 Conclusion and Recommendations

Rice is the main summer crop in Egypt. It is a cash exportable crop that provides a main source of income to the Egyptian farmers and the national economy. However, farmers used to burn the rice straw at the farm borders, violating the law that forbids such action. Burning rice straw on the field causes socio-economic negative externalities due to the generated smoke from burning the straw produced from 1.8 million feddans of rice i.e. about 0.75 million ha. It is a main cause of the probability of premature mortality and morbidity of rural and urban individuals. Chopped rice straw silage mixed with dissolved urea and molasses at 2% and 3% of straw weight, respectively, was provided as feed, at 40% of the S.E. of the daily ration and the rest was from concentrate feed mix (oil meal, yellow corn, bran and salt) to buffalo male calves for meat production. Two feed response models were estimated to compare such ration with untreated rice straw ration. The treated rice straw ration raised the marketing weight derived from the estimated feed-response model that maximizes the gross margin above the feed costs, from 384 to 518 kg live weight. Using treated rice straw as feed for buffalo calves seems more feasible to the rural communities, than using it as a soil fertilizer or in manufacturing wood and paper. While Egypt imported red meat of 963 million dollars in 2013, due to lack of sufficient feed supply, enriched rice straw silage as feed would provide an additional 80,000 tons of carcass weight from fed buffalo calves, rather than slaughtering them as rearing veal calves. The estimated extra income generated from a buffalo-fed calf reached 50% of the average annual per capita income in Egypt, with more employment opportunities for the rural communities, and preventing probable premature mortality due to burning rice straw. 92% of the Egyptian farmers holding 88% of livestock are with small farms (less than 2 ha/farm). They should be the target of a training program on how to make enriched rice straw silage

Therefore, the study recommends: (1) providing extension services to train the small farm managers on how to make such silage as feed for buffalo feeder calves, as they represent the majority farm holdings in Egypt and hold most of livestock numbers; (2) providing options for reducing the cost of collection and transportation of rice straw; and (3) intensifying information campaigns and drives regarding environmental regulations and policies as well as increasing the demand for rice straw for other recommended uses.

References

Akter, M., Asad uzzaman, M., Hossain M. M., & Asad, L. (2015). Utilization of urea and urease treated rice straw by indigenous growing cattle. *Iranian Journal of Applied Animal Science, 5*(3), 529–535.

Central bank of Egypt. (2015). Monthly bulletin, January 2015. Cairo, Egypt.

Chemjong, P. B. (1991) Economic value of urea-treated straw fed to lactating buffaloes during the dry season in Nepal. *Tropical Animal Health and Production*, 147–154.

Devendra, D., & Sevilla, C. (2002). Availability and use of feed resources in crop–animal systems in Asia. *Agricultural Systems, 71*(1–2), 59–73.

Dobermann, A., & Fairhurst, T. H. (2002, May). Rice straw management. *Better Crops International, 16*, special supplement, 1–7.

Duc Vu, D., Cuong, L. X., Dung, C. A., & Ho Hai, P. (1999). Use of urea–molasses–multi-nutrient block and urea-treated rice straw for improving dairy cattle productivity in Vietnam. *Preventive Veterinary Medicine, 38*(2–3), 187–193.

El Asheeri, A. K. (2012). Economic return of fattening baladi and buffalo calves under prevailing feeding system in Egypt. *Journal of Animal and Poultry Production, 3*(1), 21–28. Mansoura Univ.

FAOSTAT. (2015). FAO *Statistics division*, 25 January 2015, <fao.org>.

Hazell, P. B. R., Perez, N., Siam, G., & Soliman, I. (1995). *Impact of the structural adjustment program on agricultural production and resource use in Egypt*. EPTD Discussion Paper No. 10, Environment and Production Technology Division, International Food Policy Research Institute, Washington, D.C. USA.

Helal, F. I. S., Abdel Rahman, K. M., Ahmed, B. M., & Omar, S. S. (2011). Effect of feeding different levels of concentrates on buffalo calves performance, digestibility and carcass traits. *American-Eurasian Journal of Agriculture & Environmental Science, 10*(2), 186–192.

Ibrahim, M. A. R. (2012). Water buffalo for our next generation in Egypt and in the world. *Scientific Papers, Animal Science, Series D, LV*, CD-ROM ISSN 2285-5769, ISSN-L 2285-5750. University of Agronomic Sciences and Veterinary Medicine of Bucharest, Faculty of Animal Science, Faculty of Animal Science, Hungary.

Intriligator, M. (1978). *Econometric models, techniques and applications*. Englewood Cliffs, New Jersey: Prentice Hall, Inc.

Jenkins, B. M., Baxter, L. Jr, Miles, T. R., & Miles, T. R. (1999). Combustion properties of biomass. Retrieved March, 1, 1999, Copyright 1998 Elsevier Science B.V.

Launio, C. C. Asis, C. A., Manalili Jr., R. G. & Javier, E. F. (2013). Economic analysis of rice straw management alternatives and understanding farmers' choices. A Research Project Report, Published By World Fish (ICLARM)—Economy and Environment Program For Southeast Asia (EEPSEA) Philippines Office, World Fish Philippines, Country Office, Earca Bldg., College, Los Baños, Laguna 4031 Philippines.

Naik, P. K., Mendiratta, S. K., Laxmanan, V., Mehra, Usha R., & Dass, R. S. (2004). Effect of feeding ammoniated wheat straw treated with and without hydrochloric acid on meat quality and various sensory attributes of growing male buffalo calves. *Asian Australasian Journal of Animal Sciences, 17*(4), 485–490.

Nelson, R. L., Thor, P. K., & Heaton, C. R. (1980). Rice straw burning alternative policy implications. *California Agriculture*, Issue of February, 1–9.

Rahman, M. A., Alam, A. M. M. N., & Shahjalal, M. (2009). Supplementation of urea-molasses-straw based diet with different levels of concentrate for fattening of bulls. *Pakistan Journal of Biological Sciences, 12*(13), 970–975.

Soliman, I. (1995). A model for the appraisal of the environmental impacts of the projects. In *Proceedings of The Fifth International Conference on Environmental Protection is a Must*. The National Institute Of Oceanography And Fisheries, Social Development Fund, Europe-Arab Cooperation Center, Alex, April 25–27th.

Soliman, I. (2004). The role of rural women in labor and decision making for buffalo enterprise in Egyptian agriculture. In *Proceedings of 7th World Buffalo Congress, International Buffalo Federation* (Vol. 2, P777–783), Manila, Philippine.

Soliman, I. (2006). Dairy buffalo on small farm the approach towards rural development. In *Proceedings of the 5th Asian Buffalo Congress on Social Economic Contribution of Buffalo to Rural Areas*, April 18–22, 2006, Nanning, China.

Soliman, I., & Nawar, M. (1986). Feed use pattern for livestock on the Egyptian farm. In *Proceedings of the 7th Conference of Animal Production* (pp. 290–304), Organized By The Egyptian Society of Animal Production In Collaboration With the Ministry of Agriculture of Egypt, Faculty of Agriculture, Cairo University, Giza, Egypt, Held At The Egyptian International Center For Agriculture, Cairo.

Tengyun, G. (2000). Review: Treatment and utilization of crop straw and Stover in China. *Livestock Research for Rural Development, 12*(1).

Trach, N. X. (1998). The need for improved utilization of rice straw as feed for ruminants in Vietnam: An overview. *Livestock Research for Rural Development, 10*(2).

World Bank. (2013). *Development Indicators*. Wash D. C., USA: Bank Press.

Ibrahim Soliman Dr. Ibrahim Soliman is currently an emeritus professor of agricultural Economics at Dept. Agri. Econ., Zagazig University in Egypt. He got a full professorship of agricultural Economics in 1986 and was the Chairman of the Dept. Ag. Econ., ZU, (1999-2005). He worked as associate research professor at American University in Cairo (AUC), (1984-1985); an economic consultant of Kuwait Livestock and Transportation Company (KLTT), (1988-1989); a research scientist at Kuwait Institute for Scientific Research (KISR), (1989-1990); the economic consultant of the Egyptian ministry of trade and supply (1996-2003). He started his career as a teaching and research associate at Faculty of Agri., Ain Shams University 1967-1977. He got his B. Sc. of animal science from Ain Shams University in Cairo, Egypt in 1967. His first M.Sc. of animal physiology was in 1970. His second M.Sc. of agricultural economics was from Ain Shams University in 1973. His graduate studies for PhD were in both Iowa State University and Ain Shams University. He got his PhD degree in 1978. He finished all English language courses up to the advanced level at American University in Cairo (1968-1970). He was a fellow of the academy of scientific research and technology, (1991- 2004), a member of: the national specialized councils of the Egyptian presidency (1983-2011), the supreme council of culture (2011-2013, the permanent committee for promotion of universities staff in Egypt (1999 till now).

Dr. Ibrahim Soliman is currently an emeritus professor of agricultural Economics at Dept. Agri. Econ., Zagazig University in Egypt. He got a full professorship of agricultural Economics in 1986 and was the Chairman of the Dept. Ag. Econ., ZU, (1999-2005). He worked as associate research professor at American University in Cairo (AUC), (1984-1985); an economic consultant of Kuwait Livestock and Transportation Company (KLTT), (1988-1989); a research scientist at Kuwait Institute for Scientific Research (KISR), (1989-1990); the economic consultant of the Egyptian ministry of trade and supply (1996-2003). He started his career as a teaching and research associate at Faculty of Agri., Ain Shams University 1967-1977. He got his B. Sc. of animal science from Ain Shams University in Cairo, Egypt in 1967. His first M.Sc. of animal physiology was in 1970. His second M.Sc. of agricultural economics was from Ain Shams University in 1973. His graduate studies for PhD were in both Iowa State University and Ain Shams University. He got his PhD degree in 1978. He finished all English language courses up to the advanced level at American University in Cairo (1968-1970). He was a fellow of the academy of scientific research and technology, (1991- 2004), a member of: the national specialized councils of the Egyptian presidency (1983-2011), the supreme council of culture (2011-2013, the permanent committee for promotion of universities staff in Egypt (1999 till now).

He acquired the State prize of Agriculture in 1987 from Egyptian Academy of Science. The president of Egypt provided the Frist Class Medal of Excellency to Dr. Ibrahim Soliman for his excellent efforts in agricultural development. He acquired the Zagazig university prize of appreciation in agriculture sciences in 2005 and the university prize of Scientific Excellency in 2009, 2012, 2014 and 2017. He also got certificates of appreciation from several international scientific associations.

What Does the Young Generation Want to Eat and Do for Being Healthy from the Perspective of Today and the Future?

Püren Veziroğlu, Kenan Çiftçi, Bülent Miran and Ayça Nur Şahin

Abstract Being healthy includes both physical and mental conditions. It is a fact that individuals can create their mental peace and care about their body for now and the future with the help of social activities and healthy eating. 190 student were interviewed face to face via a structured questionnaire. The data was analyzed with descriptive statistics and the Analytic Hierarchy Process (AHP) that was particularly used to measure the sensitivity weights for food habits, and future choices for being healthy. Which choices to be taken into consideration depend on the maximum benefit that the students would get with respect to the Likert scores and AHP scores. Hence a 0–1 integer optimization model provides us with the necessary solution. As a result of the study, students who belong to Best #1 are male, educated in a city/big city and have higher income compared to Best #2 and Best #3. This means that it is crucial for the 41 students to eat organic and non-GMO products. They prefer to consume it now and the most important thing is that they will increase their consumption where these foods are easily accessible. Another considerable finding is that only students who belong to Best #3 prefer to consume organic foods in the future. Furthermore their income level is lower compared to the other two groups. In addition the majority of the students are female in the group. Consumers with low income tend to postpone their more expensive activities. This study is an introductory practice of BeCa Analysis. For future studies, researchers are recommended to expand the number of respondents to represent different types of decision-makers. That would be beneficial for institutions to define the consumer profile that would purchase goods or services for better and healthy living with the best satisfaction.

P. Veziroğlu (✉)
Çukurova University, Adana, Turkey
e-mail: purenveziroglu@gmail.com

K. Çiftçi
Yüzüncü Yıl University, Van, Turkey

B. Miran · A. N. Şahin
Ege University, İzmir, Turkey

© Springer International Publishing AG, part of Springer Nature 2018
K. Mattas et al. (eds.), *Sustainable Agriculture and Food Security*,
Cooperative Management, https://doi.org/10.1007/978-3-319-77122-9_3

39

Keywords Food habits · Attitude · Analytic hierarchy process
Binary integer programming

1 Introduction

Being healthy includes both physical and mental conditions. It is a fact that individuals can create their mental peace and care about their body for now and the future with the help of social activities and healthy eating. Healthy eating contributes to an overall sense of well-being, and is a cornerstone in the prevention of a frequency of conditions, including heart disease, diabetes, high blood pressure, stroke, cancer, dental caries and asthma. For children and young people, eating is particularly important for healthy growth and cognitive development (Shepherd et al. 2006). Individual-level factors related to food choices and eating behaviours include cognitions, behaviors, biological and demographic factors. These individual factors can impact on food choices through characteristics such as motivations, self-efficacy, outcome expectations, and behavioural capability. Environmental contexts related to eating behaviours include social environments, physical environments, and macro-level environments. The social environment includes interactions with family, friends, peers, and others in the community and may impact food choices through mechanisms such as role modeling, social support, and social norms. The physical environment includes the multiple settings where people eat or procure food such as the home, work sites, schools, restaurants, and supermarkets. The physical settings within the community influence which foods are available to eat and have an impact on barriers and opportunities that facilitate or hinder healthy eating. Macro-level environmental factors play a more distal and indirect role but have a substantial and powerful effect on what people eat. Macro-level factors operating within the larger society include food marketing, social norms, food production and distribution systems, agricultural policies, and economic price structures (Story et al. 2008). Previous literature has discussed eating behaviours and habits in various communities such as college students (Driskell et al. 1979, 2005, 2006; Dinger and Waigandt 1997; Garcia and Mann 2003; Navia et al. 2003; Von Ah et al. 2004; Davy et al. 2006); geriatrics (Rosenbloom and Whittington 1993; Roberts et al. 1994; Simmons et al. 2001; Donini et al. 2005); and adults (Schoenborn 1986; Johansson et al. 1999; Choe et al. 2003; Keski-Rahkonen et al. 2003; Hearty et al. 2007; King et al. 2009). Discussing the literature about college students' eating habits, it is noted that college students tend to have poor eating habits and consume fewer fruits and vegetables (Driskell et al. 2005; Racette et al. 2005). The critical time of life for food choice is when people step out independently for the first time and begin to make all of their own food decisions. For many people, this is the transition to college life (Deshpande et al. 2009). From this perspective, there is limited literature using the method of AHP. Mahmoodabad et al. (2016) discussed the healthy eating and educational necessities of female adolescents in their studies by using the method of AHP and found that seven major wrong eating habits of female adolescents are identified: junk food consumption, drinking soda and sweet drinks, eating fast food, deleting main meals, improper diets, low intake of vegetables, and not eating breakfast. Among these, low intake of vegetables, eating fast food, and not eating breakfast, with a weight rate of 32.4%, 19.4% and 19.3% respectively, are specified as the first three priorities of education. Training the students about using vegetables has the highest weight factor in the tables. Low intake of vegetables hits first place with regard to both prevalence and flexibility criteria (Mahmoodabad et al. 2016). According to the literature review study of Liberatore and Nydick (2008), it can be seen that the AHP technique is mostly used in health and medical care decisions (Liberatore and Nydick 2008).

The above-mentioned literature takes mainly into consideration the present behaviours and preferences of young people. The present study examines the preferences of the young as regards being healthy from the perspective of today and the future taking into account various characteristics and preferences. Furthermore, the study gains originality with a new methodological approach (BeCA) in which AHP scores and preferences in Likert scales are modelled via binary programming to achieve the best combination of the students' attributes.

Turkey has a young and growing population where five in one individuals falls into the 15–25 age interval (Anonymous 2016). For this reason young people's eating habits and activities in terms of preferring to live in a healthy way matters. Furthermore, young people will be the adults of the future and will raise new generations.

The aim of this study is to get the clues which will help to define policies and strategies that are best fitting for future decision-makers who are currently the youth of today. The study also aims at probing what to do in the future for a practical healthy environment oriented to today's young generation with respect to eating behaviour. The method used in the study generates the best and most likely applicable framework to the policy makers.

2 Method

In the study, a primary dataset was used. Probability Proportionate to Size (PPS) (Newbold 1988) was used to calculate the sample size. In the formula, the p value was selected as 0.50 to get the largest sample size (Miran 2002). The formula can be seen as follows:

$$n = \frac{Np(1 \quad p)}{(N-1)\sigma_{p_x}^2 + p(1-p)} \tag{1}$$

where n represents the sample size; N represents population size; p represents the proportion of students in the population and $\sigma^2_{p_x}$ is the symbol of variance. The sample size calculated by the formula is 190.

For collecting the data, the students in the sample were interviewed face to face via a structured questionnaire. The data was analyzed with descriptive statistics and the Analytic Hierarchy Process (AHP) that was particularly used to measure the sensitivity weights for food habits, and future choices for being healthy. AHP helps to find out the weights of each criterion and alternative. Weighted criteria and alternatives and some other variables representing preferences were used for modeling the best design regarding nutrition and future health. AHP was designed for the answer of the question "Which of the following is important for you to live your life in a healthy way?" For the other questions (see Table 1), a Likert scale was used.

Models that characterize respondents' approach to the health life style were determined with the help of the Binary (0–1 Integer) programming. The binary model can be shown theoretically as follows:

$$Max \sum_{i=1}^{n} y_i \tag{2}$$

$$\sum_{i=1}^{n} a_{ij}x_j + \sum_{i=1}^{n} b_{ik}z_k \geq \left(max\, a_{ij} + max\, b_{ik}\right)y_i \tag{3}$$

$$\sum_{j=1}^{m} x_j = 1 \tag{4}$$

Table 1 Attributes and levels for being healthy

Which of the following is important for you to live your life in a healthy way?	When would you prefer to consume organic foods?	What should change for increasing your organic food consumption?
•Consuming organic/non GMO products •Staying on a diet •Physical activity •Being socially active •Reading books •Working/having a job	•At present •In the future	•Income •Information •Price •Accessibility •Labeling
Which of the following would you prefer to consume as "organic"?		Which of the following items do you consume as organic?
•Fruit •Vegetables •Meat •Meat products •Milk •Fabrics		•Fruit •Vegetables •Meat •Meat products •Milk •Fabrics

$$\sum_{k=1}^{t} z_k = 1 \tag{5}$$

$$x_j : 0 - 1, z_k : 0 - 1 \tag{6}$$

where y_i is the student i (0–1); x and z represent probable selection criteria or alternatives; n represents number of students; m is the number of criteria; t is the number of choices; a represents the criterion weight; and b represents the weight of choice. Which choices to be taken into consideration depend on the maximum benefit that the students would get with respect to the Likert scores and AHP scores. Hence a 0–1 integer optimization model provides us with the necessary solution.

In Table 1, five attributes and their levels are shown. The attributes and their levels are used for "the best combinations of alternatives" (BeCa) to analyze.

3 Findings

The following tables present the profile of respondents in terms of age, location, education level and income. There are 114 respondents (60%) who are in the 21–23 age range. Out of 190, 135 respondents (71%) have completed their high school education in a city or big city. Furthermore, only 18 respondents spent their childhood in a village, while 172 respondents (90.5%) spent their childhood in a town or city or big city (See Table 2).

Table 3 shows education levels of respondents' parents; it can be said that 81% of them graduated from high school or above. From comparisons between fathers and mothers of respondents, it can be seen that only 60% of mothers graduated from high school or above.

Table 4 presents the profile of the respondents in terms of their financial status[1]. 8.9% of respondents have a monthly income below the minimum range. 47.4% of respondents were paid grants.

3.1 Best Combination of Alternatives (BeCa) Analysis

The BeCa analysis results are given below. Out of a total of 190 students only 87 (45%) students could fit into one of "the best" combinations. Table 5 shows the best

---

[1]For 2015 the minimum average wage is 950 TL. Furthermore, the 2015 average exchange rate from Euro to Turkish Lira is 3.2 TL.

Table 2 Age and location information

Age group	18–20		21–23		24 and above	
	n	%	n	%	n	%
	45	23.7	114	60.0	31	16.3
Completed high school education in	Town		City		Big city	
	n	%	n	%	n	%
	54	28.4	66	34.7	69	36.3
Childhood spent in	Village		Town		City		Big city	
	n	%	n	%	n	%	n	%
	18	9.5	49	25.8	62	32.6	61	32.1

combinations which fit into the preferences of the respondents on the issues considered. All students could not fit into these models since some of them do not have a consensus on these issues.

3.2 Discussion of Combinations

- *Best Combination 1 (Best 1): 41 students out 190 agree that*:

  For being healthy: organic/non-GMO products should be consumed at present; if organic/non-GMO products are accessible, they should increase their consumption. They prefer to consume firstly organic fruits in the future but they prefer to consume organic meat at present.

- *Best Combination 2 (Best 2): 29 out 190 students agree that*:

  For being healthy: they think they should read books, they think they should consume organic products at present, if organic/non-GMO products' price is low, they would increase their consumption. They prefer to consume firstly organic fruits in the future while they prefer to consume organic milk at present.

- *Best Combination 3 (Best 3): 17 out of 190 students agree that*:

  For being healthy: they think they should have a job, they think they should consume organic products in the future, if organic/non-GMO products are labeled, they would increase their consumption. They prefer to consume firstly organic fruits in the future while they tend to consume organic meat products at present.

Table 3 Parents' education level

	Primary School		Secondary School		High School		University		Master's Degree		Uneducated	
	n	%	n	%	n	%	n	%	n	%	n	%
Father's education level	19	10	15	7.9	74	38.9	72	37.9	8	4.2	2	1.1
Mother's education level	43	22.6	30	15.8	65	34.2	45	23.7	4	2.1	3	1.6

Table 4 Family income, pocket money and monthly income

Family income	Below 1000 TL		1000–1999 TL		2000–4999 TL		5000–9999 TL		10,000 TL and above	
	n	%	n	%	n	%	n	%	n	%
	17	8.9	78	41.1	85	44.7	9	4.7	1	0.5
Pocket money (Monthly)	Below 100 TL		100–199 TL		200–499 TL		500–999 TL		1000 TL and above	
	n	%	n	%	n	%	n	%	n	%
	16	8.4	27	14.2	77	40.5	58	30.5	12	6.3

Grants and/or income	None		Below 100 TL		100–199 TL		200–499 TL		500–999 TL		1000 TL and above	
	n	%	n	%	n	%	n	%	n	%	n	%
	24	12.6	31	16.3	14	7.4	90	47.4	24	2.1	7	3.7

Table 5 Best top three choices for maximum benefit according to the respondents' profiles

Male (27%), age interval 21-23 (14%), completed high school education in city + big city (15%), household income 2000-4999 TL (12%), monthly spending money 200- 999 (14%)

Male (9%), age interval 21-23 (7%), completed high school education in city + big city (11%), household income 1000-1999 TL (8%), monthly spending money 200-999 (10%)

Male (6%), age interval 21-23 (14%), completed high school education in city + big city (7%), household income 1000-1999 TL (5%), monthly spending money 200-999 (5%)

Best choices for maximum benefit	Best #1	Best #2	Best #3
What does she / he see as important to live a healthy life?	Organic/non-GMO products	Reading Books	Having a job
Would you prefer to consume organic foods today or in the future?	Now	Now	Future
What should change for increasing your organic food consumption?	Accessibility	Price	Labeling
Which of the following would you prefer to consume organic in the future ?	Fruits	Fruits	Fruits
Which of the following are you consuming organic?	Meat	Milk	Meat Products
Number of the students in the group	41	29	17
%	21.579	15.263	8.947
No of DMUs assigned to 3 best groups	45.79 %		

4   Conclusions

The study reveals that students who belong to Best #1 are male, educated in a city/big city and have higher income compared to Best #2 and Best #3. This means that it is crucial for the 41 students to eat organic and non-GMO products. They prefer to consume it now and the most important thing is that they will increase their consumption where these foods are easily accessible. Another considerable finding is that only students who belong to Best #3 prefer to consume organic foods in the future. Furthermore their income level is lower compared to the other two groups. In addition the majority of the students are female in the group. Consumers with low income tend to postpone their more expensive activities. This study is an introductory practice of BeCa Analysis. For future studies, researchers are recommended to expand the number of respondents to represent different types of decision-makers. That would be beneficial for institutions to define the consumer profile that would purchase goods or services for better and healthy living with the best satisfaction.

References

Anonymous. (2016). "TURKSTAT." Retrieved June 1, 2016 from http://www.tuik.gov.tr.

Choe, J., Ji, S., Paik, H., & Hong, S. (2003). A study on the eating habits and dietary consciousness of adults in urban area. *Journal of the Korean Society of Food Science and Nutrition.*

Davy, S. R., Benes, B. A., & Driskell, J. A. (2006). Sex differences in dieting trends, eating habits, and nutrition beliefs of a group of midwestern college students. *Journal of the American Dietetic Association, 106*(10), 1673–1677.

Deshpande, S., Basil, M. D., & Basil, D. Z. (2009). Factors influencing healthy eating habits among college students: An application of the health belief model. *Health Marketing Quarterly, 26*(2), 145–164.

Dinger, M. K., & Waigandt, A. (1997). Dietary intake and physical activity behaviors of male and female college students. *American Journal of Health Promotion, 11*(5), 360–362.

Donini, L. M., Savina, C., & Cannella, C. (2005). Eating habits and appetite control in the elderly: The anorexia of aging. *International Psychogeriatrics, 15*(1), 73–87.

Driskell, J. A., Keith, R. E., & Tangney, C. C. (1979). Nutritional status of white college students in Virginia. *Journal of the American Dietetic Association, 74*(1), 32–35.

Driskell, J. A., Kim, Y.-N., & Goebel, K. J. (2005). Few differences found in the typical eating and physical activity habits of lower-level and upper-level university students. *Journal of the American Dietetic Association, 105*(5), 798–801.

Driskell, J. A., Meckna, B. R., & Scales, N. E. (2006). Differences exist in the eating habits of university men and women at fast-food restaurants. *Nutrition Research, 26*(10), 524–530.

Garcia, K., & Mann, T. (2003). From 'I wish' to 'I will': Social-cognitive predictors of behavioral intentions. *Journal of Health Psychology, 8*(3), 347–360.

Hearty, A., McCarthy, S., Kearney, J., & Gibney, M. (2007). Relationship between attitudes towards healthy eating and dietary behaviour, lifestyle and demographic factors in a representative sample of Irish adults. *Appetite, 48*(1), 1–11.

Johansson, L., Thelle, D. S., Solvoll, K., Bjørneboe, G.-E. A., & Drevon, C. A. (1999). Healthy dietary habits in relation to social determinants and lifestyle factors. *British Journal of Nutrition, 81*(03), 211–220.

Keski-Rahkonen, A., Kaprio, J., Rissanen, A., Virkkunen, M., & Rose, R. J. (2003). Breakfast skipping and health-compromising behaviors in adolescents and adults. *European Journal of Clinical Nutrition, 57*(7), 842–853.

King, D. E., Mainous, A. G., Carnemolla, M., & Everett, C. J. (2009). Adherence to healthy lifestyle habits in US adults, 1988–2006. *The American Journal of Medicine, 122*(6), 528–534.

Liberatore, M. J., & Nydick, R. L. (2008). The analytic hierarchy process in medical and health care decision making: A literature review. *European Journal of Operational Research, 189*(1), 194–207.

Mahmoodabad, S. M., Mehrabbeyk, A., Mozaffari-Khosravi, H., & Fallahzadeh, H. (2016). Identifying and prioritizing educational needs of female adolescents in relation to healthy eating based on analytic hierarchy process model. *Global Journal of Health Science, 9*(2), 223.

Miran, B. (2002). *Temel istatistik*. Ege Üniversitesi Basımevi: İzmir.

Navia, B., Ortega, R. M., Requejo, A. M., Mena, M. C., Perea, J. M., & Lopez-Sobaler, A. M. (2003). Influence of the desire to lose weight on food habits, and knowledge of the characteristics of a balanced diet, in a group of Madrid university students. *European Journal of Clinical Nutrition, 57*(S1), S90–S93.

Newbold, P. (1988). *Statistics for business and economics*. Prentice Hall, ISBN 0-13-845330-6.

Racette, S. B., Deusinger, S. S., Strube, M. J., Highstein, G. R., & Deusinger, R. H. (2005). Weight changes, exercise, and dietary patterns during freshman and sophomore years of college. *Journal of American College Health, 53*(6), 245–251.

Roberts, S. B., Fuss, P., Heyman, M. B., Evans, W. J., Tsay, R., Rasmussen, H., et al. (1994). Control of food intake in older men. *JAMA, 272*(20), 1601–1606.

Rosenbloom, C. A., & Whittington, F. J. (1993). The effects of bereavement on eating behaviors and nutrient intakes in elderly widowed persons. *Journal of Gerontology, 48*(4), S223–S229.

Schoenborn, C. A. (1986). Health habits of US adults, 1985: The "Alameda 7" revisited. *Public Health Reports, 101*(6), 571.

Shepherd, J., Harden, A., Rees, R., Brunton, G., Garcia, J., Oliver, S., et al. (2006). Young people and healthy eating: A systematic review of research on barriers and facilitators. *Health Education Research, 21*(2), 239–257.

Simmons, S. F., Osterweil, D., & Schnelle, J. F. (2001). Improving food intake in nursing home residents with feeding assistance a staffing analysis. *The Journals of Gerontology Series A: Biological Sciences and Medical Sciences, 56*(12), M790–M794.

Story, M., Kaphingst, K. M., Robinson-O'Brien, R., & Glanz, K. (2008). Creating healthy food and eating environments: Policy and environmental approaches. *Annual Review of Public Health, 29*, 253–272.

Von Ah, D., Ebert, S., Ngamvitroj, A., Park, N., & Kang, D. H. (2004). Predictors of health behaviours in college students. *Journal of Advanced Nursing, 48*(5), 463–474.

Tourist's Behaviour Towards Local Cretan Food

A. Malek Hammami, John L. Stanton, Drakos Periklis,
George Baourakis, Gert van Dijk and Spyridon Mamalis

Abstract Gastronomic tourism is becoming an important part of the tourism industry. However, there are very few studies on food tourists' behaviour comparing with those about tourists' behaviour in general. The current study provides an analysis of tourists' behaviour based on the perceived value (PV) and the satisfaction (ST) to predict the intention to revisit (IN) using the Modified Theory of Reasoned Action (TRA). The purpose of this study is to unveil the relation between perceived value and intention to revisit (H1), Perceived Value and satisfaction (H2)

A. M. Hammami (✉) · D. Periklis · G. Baourakis
Mediterranean Agronomic Institute of Chania, CIHEAM-MAICh,
Alsyllio Agrokepio, 1Makedonias str Chania, Crete 73100, Greece
e-mail: hammamimalek90@gmail.com

D. Periklis
e-mail: drakos@maich.gr

G. Baourakis
e-mail: baouraki@maich.gr

J. L. Stanton
Saint Joseph's University, 5600 City Ave, Philadelphia, PA 19131, USA
e-mail: jstanton@sju.edu

D. Periklis
Department of Economics, University of Crete, Gallos Campus,
741 00 Rethymno, Greece

G. Baourakis
Center for Entrepreneurship and Stewardship, Nyenrode Business Universiteit,
Breukelen, The Netherlands

G. van Dijk
Nyenrode Business Universiteit, Straatweg 25, 3621 BG Breukelen,
P.O. Box 130, 3620 AC Breukelen, The Netherlands
e-mail: g.vdijk@nyenrode.nl

S. Mamalis
Department of Business Administration of Technological Educational Institute of Kavala,
Kavala, Greece
e-mail: mamalis@econ.auth.gr

© Springer International Publishing AG, part of Springer Nature 2018
K. Mattas et al. (eds.), *Sustainable Agriculture and Food Security*,
Cooperative Management, https://doi.org/10.1007/978-3-319-77122-9_4

and satisfaction and intention to revisit (H3) using confirmatory factor analysis as a statistical tool. Results showed that the hypotheses H1, H2 and H3 were significant ($p < 0.01$) and supported. H3 ($\beta = 0.35$) shows that satisfaction is an antecedent to the intention to revisit. The same is true for H1 ($\beta = 0.19$) and H2 ($\beta = 0.45$) which is proof that perceived value can predict intention to revisit and satisfaction respectively. The contribution of this study is intended to provide empirical results using the modified theory of reasoned action to predict tourists' behaviour toward Cretan local food which proved that local food can influence the intention to revisit. The implications will be useful for tourism managers, decision makers and destination marketing organizations in Crete and Greece.

Keywords Gastronomic tourism · Tourists' behaviour · Perceived value
Satisfaction · Intention to revisit · Confirmatory factor analysis (CFA)
Modified theory of reasoned action (TRA) · Cretan local food · Greece

1 Introduction

Tourism is considered as an important source of economic growth around the world. For many countries it is a crucial economic sector to rely on, along with agriculture and industry. Mediterranean countries having unique environment, such as Greece, consider tourism as a primordial sector.

In order for this sector to succeed, the touristic destination should be promoted to attract as much tourists as possible. Earning tourists' loyalty is also very important in tourism industry; in fact visitors who had a good stay experience in one destination would certainly have at least an increased intention to revisit it and would help in the destination promoting process to their social environment through word of mouth. Thus, the success of the touristic sector depends closely from the touristic behaviour. For that reason, many studies have been done about Tourists' behaviour, satisfaction and perception.

Meanwhile, Mediterranean civilizations are considered to be among the oldest civilizations in the history of humanity. Food and cuisine in this area of the world is distinct and more or less specific to the Mediterranean region. Besides the good sunny weather, they have, these touristic countries are relying on their long history and culture to promote tourism. Gastronomy, as a subset of cultural tourism, played a major role in the success of Mediterranean tourism generally, and in Greece especially. In fact, Cretan cuisine is a remarkable landmark of Greece as a destination. It is symbolizing the famous healthy and fresh Mediterranean cuisine from which is derived the Mediterranean diet that many nutritionists recommend it for its organic and constructive characteristics.

Tourists, arriving to Crete to enjoy the sun, the sea and landscapes, are experiencing the local Cretan gastronomy at least once a day. It is indeed a remarkable experience in the tourist's stay. But how is it seen by the gastronomic tourist?

2 Background

2.1 Tourism in Greece

Greek the tourism is a key economic activity and it is considered one of the most important sectors. During the first 8 months of 2015 for instance, Greece hosted 26 million tourists contributing with 18% to the Gross Domestic Product (GDP) (http://www.topontiki.gr 2015).

2.1.1 Gastronomic Tourism in Crete

Crete possesses a great cultural and gastronomic wealth in terms of quality and genuineness of the local products. Numerous groups including government and tour agents see a good opportunity to develop the gastronomic tourism in the island making it a gastronomic destination. And this by developing gastronomic tourism activities and services such as; seminars about Cretan cuisine, gourmets and wine conoisseurs' events, organized visits where traditional products are produced, cooking events...etc. (http://www.incrediblecrete.gr).

Wine tasting, which is a specific aspect of gastronomy, is very popular and has a long history in Crete. In fact, all the region of south Europe and especially the south Caucasus is the birthplace of "Vistis Vinifera", the wine bearing vine, several varieties that are almost exclusively cultivated in those regions. Crete is very rich in wine production, where wine tourism is flourishing. Tourists can visit the famous "wine roads of Heraklion" which are among the oldest in Crete and in the world. These roads lead tourists to age-old vineyards, historical villages, old monasteries, antiquities, contemporary cultural sites, as well as wine factories that are open to the public. Also, Crete is known by its "Liatiko", the red wine of Crete which is a DAFNES Protected Designation of Origin. Besides, there are many Wine tour companies taking visitors to wineries and special wine events open to public (http://www.incrediblecrete.gr).

2.2 Consumer Behavior

It is defined as the study of individuals and groups and the process they activate to select, secure, use and dispose of products and services in order to satisfy needs. It also includes the impact of these processes on the consumer and the society (Sabine 2012).

Consumer behavior attempts to understand the decision making process of buyers, individuals and groups. It studies the characteristics of individuals, such as demographics and behavioral variables, in order to understand the peoples 'wants (Kahle and Close 2011).

Environmental Influences		Consumer's Black Box		Consumer's Response
Marketing Factors	Environmental Factors	Individual characteristics	Decision Process	
• Product	• Economic	• Attitudes	• Problem recognition	• Product choice
• Price	• Technological	• Motivation		• Brand choice
• Place	• Political	• Perceptions	• Information search	• Dealer choice
• Promotion	• Cultural	• Personality	• Alternative evaluation	• Purchase timing
	• Demographic	• Lifestyle	• Purchase decicion	• Purchase amount
	• Natural	• Knowledge	• Post-Purchase behaviour	

Fig. 1 The black box model. (*Source* Sandhusen 2000)

To better represent the consumer behavior and his response to a product/service marketing stimuli, an adequate model is designed to present the interaction between the stimuli, consumer characteristics, the decision process and the consumer's response. As shown by Fig. 1, the Black Box Model is related to the Black Box theory where the focus is on the buyer's response to stimuli (Sandhusen 2000).

Consumer behaviour and attitude could be better understood after the purchasing act.

2.2.1 Customer Satisfaction

a. Customer satisfaction:

Customer satisfaction is the key to companies' competitiveness (Bitner and Hubbert 1994). In fact it is a main factor to retain customers, build loyalty and as a consequence create more profit (Reichheld 1996). The customer satisfaction is often defined as the post purchase comparison between the pre-purchase expectation and the received performance (Oliver 1980) or simply it is the global evaluation that customer makes after purchase (Campo and Yagüe 2009) but still "an evaluation of an emotion" by Hunt in 1977 the simplest definition.

b. Tourist satisfaction in the tourism industry:

Applying customer satisfaction on the tourism industry, which is considered as a grouped number of different industries such as travel, hospitality, food, entertainment...etc., was the concern of many empirical studies. The tourist satisfaction is a vital issue to provide managerial guidance to the tourism industry (Dmitrovic et al. 2009). For almost all the destinations, tourist satisfaction is considered as an important source of comparative advantage (Fuchs and Weiermair 2004; Buhalis 2000). Therefore, monitoring this satisfaction is helpful to managers to identify strategic objectives, prepare tactical and operational plans to increase the competitiveness and make more profit (Dmitrovic et al. 2009; Lee et al. 2008; LeHew and Wesley 2007; Turner and Reisinger 2001; Heung and Cheng 2000; Soderlund 1998; Lee et al. 2008; Hui et al. 2007; Pawitra and Tan 2003; Huang and Xiao 2000; Heung 2000; Pizam et al. 1978).

2.2.2 Perception

To understand the consumer's behavior towards one good/service, it is necessary to analyze his/her perception about that good/service and know his preference.

So the perception here is a key element in the consumer behavior. It is defined as the organization, identification and interpretation of sensory informations in order to represent and understand the environment (Schacter 2012). It must be clear that perception is not a passive receipt of these signals but shaped by learning, memory, expectation and attention (Bernstein 2010).

Concerning the perception of taste (gustation), it is the ability to perceive the flavor of substances including but not limited to food. Humans receive tastes through sensory organs called "Taste buds" or "Gustatory Calyculi" located on the upper surface of the tongue. There are five primary tastes; sweetness, bitterness, sourness, saltiness and umami. All other tastes are combinations of these basic tastes (De Vere and Calvert 2010).

- Perceived value:

Perceived value is recognized as a main factor for gaining a competitive edge for business success (Parasuraman 1997). Although it is not easy to identify the concept or to measure perceived value is a measure to examine customers' purchasing intentions the value in numbers (Parasuraman and Grewal 2000). It is defined by Zeithaml (1988) as the customer's judgment of a product based on his/her perception for what was given and what was received.

Patterson and Spreng (1997) described perceived value as a cognitive-based construct. This cognitive response leads to satisfaction which is an affective/emotional response (Cronin et al. 2000; Tarn 1999) which will predict easily the behavior intension. Woodruf (1997) has emphasized that measurement of customer satisfaction should be accompanied by the measurement of perceived value to better understand the consumer's perception. As he stated: "When triggered to make an evaluation, a customer constructs some notions, learned from past and present experiences, about what value they desire".

3 Methodology

3.1 Theoretical Overview for the Proposed Model

The attitude as a concept has been very important in tourism industry as well as in Sociology. In fact many researchers believe that attitude is the most important factor in the consumer behavior understanding (Walters 1978; Wilkie 1994). By summarizing the consumer's attitude toward a product, the marketers can have valuable informations for their product, so they can be better prepared for the marketing process (Mowen and Minor 1998). Attitudes can be useful in segmenting markets, evaluating marketing actions, and choosing target segments. In deed according to many studies, marketing success in strongly related to the understanding of how human attitude is developed and how it influences the consumer behavior. Attitude is a momentaneous and multi-dimensional concept as opposed to the uni-dimensional construct accepted in earlier studies (Loudon and Della Bitta 1988).

Wilkie (1994) assumed that human behavior is a combination of mental, emotional and physical dimensions. Later these dimensions were clearly divided into: Cognitive, affective and conative.

According to Fishbein and Ajzen (1975), the "Theory of Reasoned Action (TRA)" can be a good tool to depict consumer's intention-behaviour. Also, TRA can represent the schematic process of the three components: cognitive, affective and Behavioral (conative).

TRA is a classic persuasive model of psychology. It is also used to understand persuasive messages. It was developed in 1967 by Martin Fishbein and Icek Ajzen, derived from the theory of attitude. TRA aims to explain the relationship between attitudes and behaviour within human action. As shown in Fig. 2, This theory is used to predict individual behaviour based on pre-existing attitude and behaviour intentions (Gillmore et al. 2002).

In this regard, the purpose of this study is to identify and to analyze the relationship between these three components: Perceived Value, Satisfaction and Intention to revisit. The chosen approach is based on "Theory of Reasoned Action": TRA which was used in previous studies and researches such as (Kim et al. 2010).

3.2 Chosen Hypothesis

The following hypotheses will be tested, as indicated in Fig. 3:

- H1: Food tourists' intention to revisit can be predicted by the perceived value.
- H2: Food tourists' satisfaction can be predicted by the perceived value.
- H3: Food tourists' intention to revisit can be predicted by his satisfaction.

a. Instrument development:

The Mediterranean Agronomic Institute of Chania, in cooperation with the Technical University of Crete and the Chania Hoteliers Association, conducted a research regarding the profile, satisfaction of services, consumption habits and budget spent of the visitors to the Prefecture of Chania. For this purpose a survey was prepared to be completed by tourists in the airport of Chania in the touristic

Fig. 2 Three-component views of attitudes (Wilkie 1994)

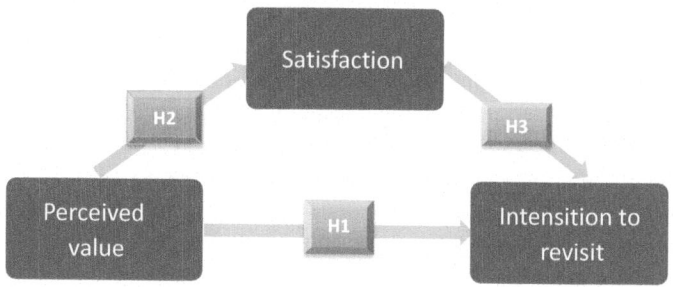

Fig. 3 Hypotheses on the proposed model

season 2015. The survey is entitled "Survey on Touristic attitude and consumer behaviour" and is made of two main parts: demographic questions and a 19 diverse questions about the destination such us: the way holidays were spent, satisfaction, accommodation, food, prices, priorities during the stay, rating the destination and finally the intention to revisit.

In order to have a better reliability and better results, some transformations have been occurred on the variables. The scaled variables were recoded into nominal just to fit the analysis using mainly the "Net Promoter Score" theory.

b. Sample and data collection:

The sample was collected in Chania International Airport. Questionnaires were given to tourists who are about to leave Chania at the end of their stay. Data collection was spread from the month of May till the month of October; during the five most crowded months of the touristic season.

4 Results and Discussion

4.1 Factor Analysis Results

Most of the literature such as "Kim et al. (2011)", Brown (2006), Kline (2005), Loehlin (2004) etc. Has shown that the CFA process is going through "Explanatory Factor Analysis" (EFA). The EFA verifies the data, divide and differentiate the variables into factors with extractions and rotations methods. The reliability (Cronbach's Alpha) of factors is also determined, just like the EFA.

The CFA specifies a model indicating which variables loading on which factors and correlations between factors. A measure of model fit is also obtained

a. Exploratory Factor Analysis (EFA) results:

The EFA was run with 6 chosen variables from the questionnaire related to the food tourism. Variables that would express both food tourism and one of the potential factors: perceived value, satisfaction or intention to revisit. Two variables were chosen for each potential factor.

The extraction method that showed the best results was: the "Principal component analysis". And the rotation method was the "Promax with Keizer normalization".

As shown in Table 1, the "Kaiser-Meyer-Olkin Measure of Sampling Adequacy" is 0.6 which is not so good but acceptable to show a medium adequacy in the identity correlation matrix. Only three factors in the initial solution have eigenvalues greater than 1. Together, they account for almost 73% of the variability in the original variables.

The variables are clearly divided into 3 main factors in this Pattern matrix; the variables were divided like it was expected in the methodology. The coefficients are good (>0.7).

- Reliability

The reliability test for this study is showing an acceptable Cronbach Alpha for the "satisfaction" factor (0.8), while less than desired numbers for the other two factors (Perceived Value and Intention to revisit) which showed values of 0.5 and 0.6 respectively.

b. Confirmatory Factor Analysis (CFA):

The CFA results were significant as shown below: Chi-square: 8,566; Degrees of freedom: 6; probability level: 0.199. The departure of the data from the model is significant at the 0.10 level.

The "Normed fit index" (NFI) has a value of 0.998—very close to 1—so the model fit is very good. The CMIN/DF ratio is set to the minimum discrepancy that if it is >2.00, it represents an inadequate fit." (Byrne 1989, p. 55). for this study, CMIN/DF = 1.428 which is considered acceptable. The "Comparative fit index"

Table 1 Pattern matrix

Variable/question	Component		
	ST	PV	IN
ST1: To what level are you satisfied with cafe' shops?	0.916		
ST2: To what level are you satisfied with taverns/restaurants?	0.885		
PV2: Please try to rate the importance of prices/cost of services that influenced you to choose this part of Crete as a touristic destination		0.893	
PV1: Please try to rate the importance of local food that influenced you to choose this part of Crete as a touristic destination		0.756	
IN1: Would you like to visit Chania/crete again?			0.827
IN2: Will you suggest it?			0.795
Cronbach Alpha	0.8	0.5	0.6
KMO	0.6		

(CFI) index is representing "The comparative fit index" and it shows 0.999 which is an excellent value since CFI values close to 1 indicate a very good fit.

"Root Mean Square Error of Approximation" (RMSEA) is 0.010. The value is considered very good since every RMSEA of about 0.05 or less would indicate a close fit of the model in relation to the degrees of freedom. Also, the PCLOSE: which is a "p-value" for testing the null hypothesis (that RMSEA is not greater than 0.05 in the population). Here PCLOSE = 1, shows a good fit since every PCLOSE > 0.5 means a good fit.

Table 2 shows the correlation and the covariance between the factors. All correlations and covariance are positive and the P-value is significant. In other words, the covariance between all factors is significantly different from zero at the 0.001 level (two-tailed).

Figure 4 is showing the correlation between factors and the standardized regression weights estimations of the variables relating them to factors.

There is a positive correlation between all the factors.

4.2 Discussions

The current study's data were collected in a purpose of discovering the profile of Crete's Tourists and their behaviour in order to predict their attitude which will help promote tourism in Crete.

Table 2 Factors covariance and correlations: (Group number 1—default model)

			Correlations	Covariance	S.E.	C.R.	P	Label
Satisfaction	<–>	Perceived	0.445	0.109	0.009	12,325	***	
Intention	<–>	Satisfaction	0.349	0.049	0.004	13,070	***	
Intention	<–>	Perceived	0.194	0.018	0.003	6,896	***	

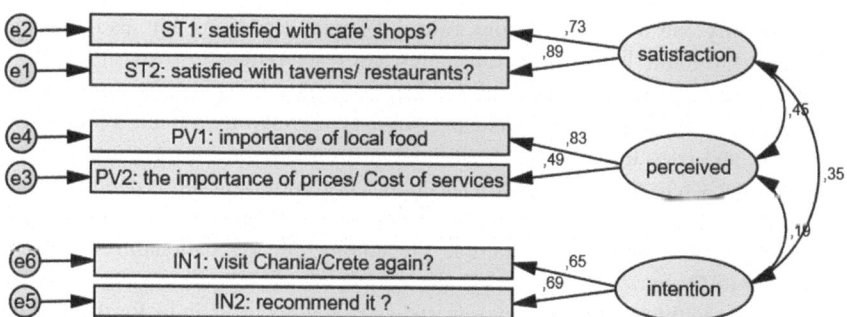

Fig. 4 Structural equation modeling (SEM). Model fit statistics: $\chi^2 = 8{,}566$, df = 6, p-value = 0.000, RMSEA = 0.01, CFI = 0.999

During the factor analysis, six variables were chosen and spread on 3 factors (Perceived Value, Satisfaction and Intention to revisit), two variables for each one.

The Exploratory factor Analysis showed a good adequacy. It regrouped the six variables into three main groups as it was expected. The reliability of the factors (Cronbach Alpha) was not as great as it should be but medium to weak.

The Confirmatory Factor Analysis showed expected results, as there is a positive correlation between the tourist/consumer's perception and satisfaction with local food and his willing to revisit the destination. The model fit was satisfactory, with reliable indicators. This study showed that local food consumption is very important in Tourists' journey. Generally, 75% of his budget will be spent on food. Tourism deciders in Greece and in Crete especially should focus on developing strategies concerning Tourists' gastronomy. The more tourists' are satisfied and developing a good image about Cretan food (perception) the more likely they will revisit the destination. It is very important to improve the image that tourists' have about local food and to build for it a solid reputation worldwide; because this will not only increase their chances to revisit but also will increase word of mouth and push other tourists to discover what they heard about. Of course the more there is an intention to revisit the better is for the tourism industry. And this is how the local food can have an important effect on success of tourism in one destination.

The previous results can provide ideas about improving Crete's potential as a tourist destination. Certainly, knowing the tourists profile (category) and anticipating their behaviour will help the deciders and tourism's managers to adjust strategies in order to explore the full potential of this destination.

The results of this study can provide guidance for strategies to improve tourist visitation rates:

- Since prices may seem more or less cheap to tourists, that are qualifying the gastronomic experience as more important than prices, the level of prices can be raised to correspond to consumers expectations but shouldn't exceed it.
- Organize more Gastronomic activities with a cultural orientation to spread and promote Cretan cuisine all over the world and encourage local food businesses. Activities like wine tasting, local cooking competitions, organized visits to local farms and traditional crafts will affect tourists' perception about local Cretan gastronomy and make them more aware of Cretan traditions, but also promote local gastronomic business such as cafés and restaurants.
- Rely on local gastronomy as a cultural Tourism, since according to numbers it is having a direct effect on increasing tourists' loyalty to the destination, by helping and encouraging local traditional gastronomic businesses.
- Promoting Crete, its culture and especially Gastronomy in those countries that showed a good attendance and responsiveness to Crete as a summer destination. Scandinavian countries for example are among the top nationalities that are arriving to Crete.
- According to this study's model, tourists' satisfaction is a main factor that can be improved by Cretan authorities and that is having a direct effect on the loyalty of tourists and on their perception about Crete as a touristic-gastronomic destination.

5 Conclusion

The market place of tourism is very competitive. A constant improvement of potentials and marketing strategies is required. The key factor to success would be to attract new visitors by promoting the destination while maintaining the loyalty of actual tourists making them revisit the destination. In order to achieve that, a study and analysis of Tourists behaviour towards a destination should be done.

Destination's visitors are always confronted to a gastronomic experience, during which the tourist is discovering other cultures through its cuisine and local food. Gastronomic tourism has been lately an important subject of studies due to its critical role into the touristic activity.

Few limitations may affect this study's results. First of all, data were collected during summer 2015. Although touristic seasons seem similar of several previous years, this study may be accurate and specific only on the 2015 season that may have some difference with previous and next seasons. For more general study a longer time series is required, for example a 10 years' data would be more representative for a more general overview. Second of all, the questionnaires were made in Chania airport only which is the second biggest airport in Crete after Heraklion. So a bigger data spread on the two airports would be more significant if we wanted to speak about Crete in general. Finally, the questionnaires were about general touristic profile and behaviour—not about food specifically—from which questions about food were collected. A more specific questionnaire only about gastronomic tourists' behaviour would be more helpful and would provide more data on food tourists especially.

References

Bernstein, D. A. (2010). *Essentials of Psychology. Cengage Learning, 123–124.* ISBN 978-0-495-90693-3.

Brown, T. A. (2006). *Confirmatory factor analysis for applied research.* New York: Guilford Press.

Bitner, M. J., Hubbert, A. R. (1994). Encounter satisfaction versus overall satisfaction versus quality, in Rust, R. T., Oliver, R. L. (Eds), *Service Quality: New Directions in Theory and Practice,* USA, Sage Publications, 72–94.

Buhalis, D. (2000). Marketing the competitive destination of the future. *Tourism Management, 21* (1), 97–116.

Byrne, B. M. (1989). Multigroup comparisons and the assumption of equivalent construct validity across groups: Methodological and substantive issues, *Multivariate Behavioral Research, 24* (4), 503–523.

Campo, S., & Yague, M. J. (2009). Exploring non-linear effects of determinants on tourists' satisfaction. *International Journal of Culture, Tourism and Hospitality Research, 3*(2), 127–138.

Chiras, D. D. (2004). *Environmental science: Creating a sustainable future.* Jones & Bartlett Learning.

Cronin, J. J., Jr., Brady, M. K., & Hult, G. T. M. (2000). Assessing the effects of quality, value, and customer satisfaction on consumer behavioral intentions in service environments. *Journal of Retailing, 76*(2), 193–201.

DeVere, R., Calvert, M. (2010). *Navigating smell and taste disorders*. Demos Medical Publishing. pp. 33–37. ISBN 978-1-932603-96-5. Retrieved March 26, 2011.

Dmitrovic, T., Cvelbar, K. L., Kolar, T., Brencic, M. M., Ograjensek, I., & Zabkar, V. (2009). Conceptualization tourist satisfaction at the destination level. *International Journal of Culture, Tourism and Hospitality Research, 3*(2), 116–126.

Fishbein, M., & Ajzen, I. (1975). *Belief, attitude, intention, and behavior*. Reading, MA: Addison-Wesley.

Fuchs, M., & Weiermair, K. (2004). Destination benchmarking: An indicator system's potential for exploring guest satisfaction. *Journal of Travel Research, 42*, 212–225.

Gillmore, M. R., Archibald, M., Morrison, D., Wilsdon, A., Wells, E., Hoppe, M., et al. (2002). Teen sexual behavior: Applicability of the theory of reasoned action. *Journal of Marriage and Family, 64*.

Heung, V. C., & Cheng, E. (2000). Assessing tourists' satisfaction with shopping in the Hong Kong special administrative region of China. *Journal of Travel Research, 38*(4), 396–404.

Heung, V. C. S. (2000). Satisfaction levels of mainland Chinese travelers with Hong Kong hotel services. *International Journal of Contemporary Hospitality Management, 12*(5), 308–315.

http://www.incrediblecrete.gr.

http://www.gesaky.com/retailer-area.

http://www.topontiki.gr. (2015).

Huang, A., & Xiao, H. (2000). Leisure-based tourist behavior: A case study of Changchun. *International Journal of Contemporary Hospitality Management, 12*(3), 210–214.

Hui, T. K., Wan, D., & Ho, A. (2007). Tourists' satisfaction, recommendation and revisiting Singapore. *Tourism management, 28*(4), 965–975.

Kahle, L. R., & Close, A. G. (2011). *Consumer behavior knowledge for effective sports and event marketing*. New York: Routledge. ISBN 978-0-415-87358-1.

Kim, Y. H., et al. (2011). *Tourism Management, 32*, 1159–1165.

Kim, Y. H., Goh, B. K., & Yuan, J. (2010). A development of a multi-dimensional scale for measuring the motivation factors of food tourists at a food event: What does motivate people to travel? *Journal of Quality Assurance in Hospitality and Tourism, 11*(1), 56–71.

Kline, R. B. (2005). *Principles and practice of structural equation modeling* (2nd ed.). New York: Guilford.

Lee, C. L., Yang, S. C., & Lo, H. Y. (2008). Customer satisfaction and customer characteristic in festival activity. *International Journal of Culture, Tourism and Hospitality Research, 2*(3), 234–249.

LeHew, M. L. A., & Wesley, S. C. (2007). Tourist shoppers' satisfaction with regional shopping mall experiences. *International Journal of Culture, Tourism and Hospitality Research, 1*(1), 82–96.

Loehlin, John C. (2004). *Latent variable models: An introduction to factor, path, and structural analysis* (4th ed.). Mahwah NJ: Erlbaum.

Loudon, D., & Della Bitta, A. J. (1988). *Consumer behavior: Concepts and applications*. New York: McGraw-Hill.

Mowen, J. C., & Minor, M. (1998). *Consumer behavior* (5th ed.). Upper Saddle River, NJ: Prentice-Hall.

Oliver, R. L. (1980). A cognitive model of the antecedents and consequences of satisfaction decisions. *Journal of Marketing Research, 17*(4), 460.

Parasuraman, A. (1997). Reflections on gaining competitive advantage through customer value. *Journal of the Academy of Marketing Science, 25*(2), 154–161.

Parasuraman, A., & Grewal, D. (2000). The impact of technology on the quality value-loyalty chain: A research agenda. *Journal of the Academy of Marketing Science, 28*(1), 168–174.

Patterson, P. G., & Spreng, R. A. (1997). Modeling the relationship between perceived value, satisfaction and repurchase intentions in a business-to-business, services context: An empirical examination. *International Journal of Service Industry Management, 8*(5), 414–434.

Pawitra, T. A., & Tan, K. C. (2003), Tourist satisfaction in Singapore–A perspective from Indonesian tourists. *13*(5), 399–411.

Pizam, A., Neumann, Y., & Reichel, A. (1978). Dimensions of tourist satisfaction with a destination area. *Annals of Tourism Research, 5*(3), 314–332.

Reichheld, F. F., Teal, T., & Smith, D. K. (1996). *The loyalty effect* (Vol. 1, No. 3, pp. 78–84). Boston, MA: Harvard business school press.

Sabine, K. (2012). *MKT 301: Strategic marketing & marketing in specific industry contexts* (p. 110). University of Mannheim.

Sandhusen, R. L. (2000). Barons‖ s Marketing–A true-to-life hypothetical company presented Business Review Books.

Schacter, D. L. (2012). Adaptive constructive processes and the future of memory. *American Psychologist, 67*(8), 603–613.

Soderlund, M. (1998). Customer satisfaction and its consequences on customer behavior revisited. The Impact of different levels of satisfaction on word-of-mouth, feedback to the supplier and loyalty. *International Journal of Service Industry Management, 9*(2), 169.

Tarn, J. L. (1999). The effects of service quality, perceived value and customer satisfaction on behavioral intentions. *Journal of Hospitality & Leisure Marketing, 6*(4), 31–43.

Turner, L. W., & Reisinger, Y. (2001). Shopping satisfaction for domestic tourists. *Journal of Retailing and consumer services, 8*(1), 15–27.

Walters, C. G. (1978). *Consumer behavior: Theory and practice* (3rd ed.). Homewood, IL: Richard D. Irwin.

Wilkie, W. L. (1994). *Consumer behavior* (3rd ed.). New York: Wiley.

Woodruff, R. B. (1997). Customer value: The next source for competitive advantage. *Journal of the Academy of Marketing Sciences, 25*(2), 139–153.

Zeithaml, V. A. (1988). Consumer perceptions of price, quality and value: A meansend model and synthesis of evidence. *Journal of Marketing, 52*(3), 2–22.

Exploring-Valuing Alternative Distribution Channels: A Systematic Literature Review of the Agrifood Sector

Kallirroi Nikolaou, Foivos Anastasiadis, Efthimia Tsakiridou and Konstadinos Mattas

Abstract The main objective of this research is to determine and describe what constitutes an alternative distribution channel for raw food products by exploring the relative literature and the practices. Alternative food channels cover an extend of channels that are depending on the local conditions and the type of agricultural distributed products. Several factors influence consumers attitudes towards alternative charnels and they are recorded. Generally consumers tend to show a preference towards alternative charnels over the last years particularly for fresh fruits and vegetables.

Keywords Alternative distribution channels · Fruits and vegetables
Consumer behavior · Local markets

Globalization and recent economic trends have created highly complex supply chains and as a result their design, organization, interactions, competencies, capabilities and management have become key issues (Ashby et al. 2012). A close study of past research has shown only some traces of a structured approach to supply chains including their weak aspects and the risks involved (Svensson 2000; Sheffi 2001; Zsidisin et al. 2000; Guertler and Spinler 2015; Dekker et al. 2013; Cantor et al. 2014; Heckmann et al. 2015). In order to have an integrated supply chain, a totally new approach needs to be implemented, whose development involves other related disciplines, such as market research and operational strategic management, incorporating empirical research theories and methodologies (Cheng and Grimm 2006; Wisner 2003). The aim of any commercial operation, including the agrifood sector, is obviously competitive advantage, which can be created by synchronizing

K. Nikolaou · E. Tsakiridou · K. Mattas (✉)
School of Agriculture, Department of Agricultural Economics, Aristotle University of Thessaloniki, Thessaloniki, Greece
e-mail: mattas@auth.gr

F. Anastasiadis
School of Engineering, Industrial Management Division of the Mechanical Engineering Department, Aristotle University of Thessaloniki, Thessaloniki, Greece

© Springer International Publishing AG, part of Springer Nature 2018
K. Mattas et al. (eds.), *Sustainable Agriculture and Food Security*,
Cooperative Management, https://doi.org/10.1007/978-3-319-77122-9_5

supply chain strategy with competitive policy (Porter 1985). This can be accomplished by establishing a wide spectrum of alternative and opportunity networks, such as distribution channels, that form a coordinated, integrated whole (Achrol 1997; Tsang 2000).

Important components of alternative distribution channels among others, are high quality, high standards and consumer-producer trust (Whatmore 2003). A customer-centered approach (Spiller 2008) and a short distribution channel direct product provision from producer to consumer, are fundamental to optimal distribution. It was found that short supply chains incorporate farmers' markets, street stalls and street markets, direct farm sales and more recently the Internet. A key priority of agriculture and rural development is to strengthen both the means of distribution and the processes and functions of the short supply chain (Burt and Wolfley 2009; Mauleón 2003; Falguera et al. 2012). A competitive perishable food industry, can not only provide healthy and safe food to consumers but also may constitute a factor in stabilizing the economy by generating jobs, for instance, even during the global economic crisis (Mattas and Tsakiridou 2010). Clearly, in the Europe of the new millennium, both consumption and production have been affected by the consumer shift to safe products of high quality (Dimara and Skuras 2005).

In the present paper, an extensive literature review was conducted in the attempt to provide a comprehensive definition as to what constitutes alternative distribution means and channels in general, and alternative agricultural distribution channels, in particular. Although numerous studies have been conducted on alternative types of distribution channels, there is no established definition of the term 'alternative'. Through the description, analysis and comparison of the articles under review it became evident that not only is this a broad field of study but also a general one in that it entails information and parameters that fit an array of cases each with its particular characteristics. The paper comprises of four main sections. More specifically, the first part is the introduction, which presents the study objectives, including a review of the literature, followed by the second section which includes the field of study on alternative agrifood sector distribution channels in relation to relevant research and the methodology. Next is the results and discussion, which examines the study findings, and finally the conclusion, giving some concluding remarks. The reason why agricultural products were focused on is because it is both a timely topic and an area open to further study. Finally, the contribution of the present study is to shed light on the constituents of alternative agricultural distribution channels, whose constructive information can be applied effectively each for their own purpose by producers, entrepreneurs, scientists, policy makers, as well as consumers.

1 Field of Study

Following is presented an account of the field of study and the methodological approach to the literature review. Firstly, in the field of study, we present key concepts regarding alternative distribution channels based on previous studies, unveiling in this way the motivation of the current study. Secondly, we provide details of the methodology employed, a systematic literature review with meta-synthesis, such as, criteria of selecting articles.

To begin with, the salient characteristics of innovativeness which reinvigorate supply chain management have appeared in conceptual and empirical studies (Chapman et al. 2003; Roy et al. 2004; Soosay et al. 2008; Panayides and Venus Lun 2009), while Yu et al. (2014) stress that integrated green supply chain management has a positive result on operational performance. In their explanation on the historic evolution of Decision Theory in management, French et al. (2009) state that for an up-to-date, successful decision-making process, the characteristic of sustainability is needed.

Sustainability, a component of sustainable development, involves meeting the needs of the present without compromising the ability of future generations to meet their own needs and by extension, sustainable distribution concerns the way of transporting goods with minimum environmental impact (Veatch and Maren 2007). Many studies have been conducted on sustainability in general, however, those related to agriculture indicatively include the following. Sydorovych and Wossink (2008), studied the meaning of agricultural sustainability, whereas the study by Borghi et al. (2014), was on sustainability in the case of agricultural products in the supply chain, and Komatsu et al. (2005) conducted an evaluation of agricultural sustainability. Heberling et al. (2012) study on the meaning of Green Net Regional Product and Meier et al. (2015) on environmental impacts of organic and conventional agricultural products furthermore Ladhari and Tchetgna (2015) examine the influence of personal values on Fair Trade consumption. On the other hand, there is an increasing number of researchers who have started providing more in-depth analysis on customer attitudes (Baxter 2012; Ellis et al. 2012). According to Schiele (2008) one reason why customers are turning to alternative ways of obtaining agricultural products is as a result of the diminishing number of individual suppliers in current supply markets. Marketers worldwide that have, on the one hand, been applying McCarthy's four Ps (product, place, price, promotion) and, on the other, consumer oriented Lauterborn's four Cs (consumer, convenience, cost, communication) (Lauterborn 1990), have been effectively competitive, as this marketing mix (Borden 1965) determines activities that are associated with defining alternative distribution channels.

Consumer interest, on the other hand, focuses on systems and processes that take the characteristics of both the products and the techniques used to produce them into consideration (Chinnici et al. 2002). As shown in a study by Xue et al. (2014), the alternative channel structures are affected by supply chain, and consumer behavior is affected by alternative channel structures. It is useful, therefore, for all aspects and components of the distribution channels to be taken into consideration in order to assure the best outcome for both producers and consumers.

In their study on the role of procurement, Pereira et al. (2009) point to the fact that there are many studies which have researched the connection between buyers, suppliers and consumers, (Essig and Amann 2009; Kovacs et al. 2008). In fact, an increasing number of researchers have started providing more in-depth analysis on customer attitudes (Baxter 2012; Ellis et al. 2012). As regards consumer behavior, usually it is influenced by both the personal needs of each individual and the external factors of the environment in which it operates. The recent global economic crisis is one such external factor that has had a significant impact on consumer behavior (Solomon 2009). In fact, although the present study, for practical purposes does not deal with the global crisis per se, the direct and indirect impact on both producer and consumer behavior during the study time period cannot be denied.

It appears that there is a growing number of studies that focus on supplier involvement and relational reliability (Ellis et al. 2012), as well as supplier development and social capital (Blonska 2010). Little or no reference, however, has been made to the impact that alternative distribution channels have had on the agriculture products market. On the other hand, there has been emphasis on 'alternative' in studies related to: agrifood products (Leyshon 2005; Leyshon et al. 2003); organic food markets (Sheng et al. 2009); consumption (Leyshon 2005; Leyshon et al. 2003); food provision systems (Watts 2003), economic geographies of exchange and circulation (Hughes 2005); and circulation of fruits and vegetables (Bao et al. 2012).

The study of the above literature review has revealed that the term 'alternative distribution channel' is highly variable. Applying the markers of time and space, however, gives a more defined orientation as to what constitutes 'alternative'. Taking the equivocal case of e-commerce for example, regarding time, it started as an alternative distribution channel for all products around the beginning of the decade of the 1990s, with the widespread commercial use of the Internet, and becoming more mainstream by 2000, when agricultural products were also introduced. This, of course, applies primarily to the developed economies of the West, such as the USA and Western Europe. It is still regarded as alternative in developing economies around the globe. Similarly, space is another determiner of alternative distribution channels, especially as regards the agrifood sector. Again, the physical size of the country, its borders, geographical landscape etc., all play a role in facilitating or not the alternative distribution channel. For perishable food products, e-commerce is efficient when the factor of 'space' is easily accessible and carries the required infrastructure. There are many regions, for example, whose terrain is difficult and there is a poor road network or insufficient technological backup. There is a growing body of research on online distribution of goods and services. Research by Shu et al. (2007) on technology and infrastructure showed the required online exchanges in Chinese agriculture, whereas, Bao et al. (2012) examined supply chain management as this is supported by the e-Commerce Service Platform for Agreement regarding the circulation of fruits and vegetables, and Jamaluddin (2013) researched the adoption of e-commerce practices among Indian farmers.

From all the above, our field of research, which is to define the term 'alternative distribution channel' required that we first give a thorough outline of the studies that have been conducted in the last two decades, both in the wider fields of economics and agriculture related to distribution channels generally and more specifically to 'alternative'. Based on the key concepts presented previously, the subject of alternative distribution channels is both important and complicated. Also, there is no clear definition, thus, it is the scope of the present paper to provide a detailed review of this extensive body of literature in the attempt to form a concise definition of 'alternative' in this case.

2 Methodology

A literature review is "a systematic, explicit, and reproducible design for identifying, evaluating, and interpreting the existing body of recorded documents" as defined by Fink (2009: 2). This ensures a higher reliability in the results and subsequent conclusions that are drawn. For a more comprehensive exploration of the study issue, a process of research synthesis is applied, which summarizes and integrates the various papers on the relevant topic (Tranfield et al. 2003). The information obtained from reading an article in isolation produces different results and/or knowledge (Denyer et al. 2009). A systematic literature review was considered to be the most appropriate methodology for our research as it has an approach that is both high in objectivity and transparency in research synthesis, whose aim is to reduce bias to a minimum (Ashby et al. 2012; Lawson et al. 2006; Jain 2010) and for this reason we considered the combination of the papers' information necessary to achieve the most comprehensive definition possible in such a novel field.

In order to access the related papers, the following key words and phrases were selected:

- Words: alternative, local/regional, green, eco-friendly, organic farming, sustainability, e-commerce, technology agriculture, agricultural marketing, fair trade, smart agriculture, short supply chain.
- Phrases: alternative agricultural distribution channels, alternative distribution networks, alternative food systems, agro/agri-food distribution channels, agro/agri-food distribution networks, new agricultural distribution channels, new distribution networks, green agricultural distribution channels, green distribution networks, environmentally-friendly distribution channels, environmentally-friendly distribution networks, local/regional distribution channels, "local distribution networks, sustainable agricultural distribution channels, sustainable distribution networks, online agricultural distribution channels, networks' applications in agriculture, agricultural transmissions lines, alternative agricultural trade, sustainable food consumption, future agricultural channels, smart distribution systems.

Since qualitative data is useful for providing interpretation of a phenomenon, the methodology was based on a search of key words and phrases that were contained in the common themes and recurring ideas in the reviewed articles in order to strengthen the evidence for an interpretation of what constitutes alternative distribution channels.

The present research was based on every possible combination of the above topics. A meta-search engine (MetaLib) was employed that had access and compared the most well-known academic databases (ABI/INFORM Global, Academic Research Library, Academic Search Premier, Emerald Journals, ISI Web of Knowledge). The first phase generated a total of 820 papers published within the last 20 years. The reduction process was based on the following criteria: duplicated results were deleted, and only articles referring to agricultural products were selected. Impact Factor is one measure of the relative importance of a journal, individual article, or researcher to literature and research (EASE), 2007, so if we base on journals with high impact factor the reliability could create more solid research. Following this, the abstracts of 240 articles were read focusing on any term related to 'alternative distribution channels'. This procedure resulted in 94 articles from 64 journals; more specific, 15 articles with a range of impact factor greater than 0.299 to 1.135 and 79 articles with a range of impact factor greater than 1.268 to 4.652 (Table 1).

In addition, Table 2 presents in more detail the names of the journals. In order to minimize the subjectivity of which articles were included and which excluded, all the authors took part in this phase of the systematic review, as Tranfield et al. propose (2003). After deleting the redundant results the findings of these papers were compared, in order to designate what constitutes agricultural product distribution channels.

For the analysis, as recommended by Denyer et al. (2009) we did not just present a simple description of the articles. The information that was gathered from the different papers on the subject of alternative agricultural distribution channels was developed into distinct categories in the attempt to define the term 'alternative distribution channel'. A combinatory and comparative methodology was applied, which allows for a 'meta-synthesis' approach of interpretation to be undertaken which takes into account the similarities and differences in the explanation of the studied topic (Tranfield et al. 2003; Burgess et al. 2006; Strauss, 1998). According to Schreiber (1997: 314) meta-synthesis "is bringing together and breaking down of findings, examining them, discovering essential features and, in some way, combining phenomena into a transformed whole". The distinct categories that arose

Table 1 Papers within impact factor range of journals	Impact factor (range)	Journals	No. of papers
	0.299–0.995	12	12
	1.028–1.135	3	3
	1.268–1.909	24	35
	2.036–2.916	16	27
	3.188–3.590	4	8
	4.074–4.651	5	9

Table 2 Journals reviewed and impact factor[a]

Journal title	Impact factor	No. of papers
Academy of Management Journal	4.074	1
African Journal of Business Management	0.299	1
Agricultural Sciences in China	0.820	1
Agricultural Systems	2.453	1
Agriculture and Human Values	1.359	1
Agriculture, Ecosystems and Environment	3.203	1
Antipode	1.885	1
Appetite	2.691	1
British Food Journal	0.649	3
British Journal of Management	1.909	1
Business Strategy and the Environment	2.877	1
Clothing and Textiles Research Journal	0.476	1
Computers & Industrial Engineering	1.690	1
Computers and Electronics in Agriculture	1.486	1
Electronic Commerce Research and Applications	1.304	2
Environment and Planning A	1.890	3
European Journal of Development Research	0.564	1
European Journal of Operational Research	1.843	2
European Management Journal	0.817	1
Food Control	2.819	2
Food Quality and Preference	2.779	1
Futures	0.995	1
Geoforum	1.877	1
Human Organization	0.56	1
Industrial Crops and Products	2.837	1
Industrial Marketing Management	1.897	2
International Food and Agribusiness Management Review	0.545	1
International Journal of Consumer Studies	0.718	1
International Journal of Information Management	2.042	1
International Journal of Logistics Management	1.135	1
International Journal of Operations & Production Management	1.518	3
International Journal of Physical Distribution & Logistics Management	1.759	3
International Journal of Production Economics	2.081	3
International Journal of Production Research	1.46	1
Internet Research	1.638	1
Journal of Business Logistics	2.886	2
Journal of Business Research	1.306	1
Journal of Cleaner Production	3.590	2
Journal of Environmental Management	3.188	3

(continued)

Table 2 (continued)

Journal of Environmental Policy & Planning	1.279	1
Journal of Happiness Studies	1.772	1
Journal of Operations Management	4.478	1
Journal of Purchasing and Supply Management	1.609	2
Journal of Retailing and Consumer Services	1.364	2
Journal of Rural Studies	2.036	6
Journal of the Academy of Marketing Science	3.41	2
Land Use Policy	2.631	1
Management Accounting Research	2.125	1
Omega	4.376	2
Organization & Environment	1.386	1
Organization Studies	2.504	1
Politics & Society	1.268	1
Progress in Human Geography	4.394	2
Sociologia Ruralis	1.362	1
Supply Chain Management:An international journal	2.916	4
Sustainability	1.077	1
The ANNALS of the American Academy of Political and Social Science	1.028	1
Transportation Research Part E: Logistics and Transportation Review	2.676	1
Trends in Food Science & Technology	4.651	3
World Development	1.733	2

[a]In alphabetical order

from the analysis are the following: 'Green', 'e-commerce', 'local food networks' (farmers' markets, food cooperatives) and the 'Community' (fair trade).

3 Results and Discussion

Concerning the field of results we attempted to create themes and categories based on their common characteristics. First of all, two general categories were formed: "established trend" and "new trend". By the term "established trend" we want to distinguish that while these channels are alternative, they have become more widespread and accepted in their usage. On the other hand, by "new trend" we mean that these two distribution channels form a relatively recent tendency that caters to society's contemporary needs. Afterwards, four such themes were revealed as a result of the meta-synthesis of the keywords and phrases investigation. Even though the interpretation of these four forms is broad, they tend to overlap in some cases. More specifically, the four forms are: 'Green', 'e-Commerce', 'Local food networks' and 'the Community'.

3.1 'Green' Channel

'Green' is a characteristic of an alternative distribution channel due to the fact that it comprises production and distribution processes that are organic, and ecologically friendly, whose express aim is the protection of the environment. Contrastingly, conventional methods focus on economic and functional distribution factors (Elith and Leathwick 2009). The growing interest in alternative food systems within the context of environmental and social sustainability (Cleveland et al. 2014), has led to reduced costs, increased efficiency, the creation of 'green' brand products, as well as a decrease in waste and pollution (Cordano 1993). Through the convergence of supply chain sustainability, interest in environmental management and operation has shifted from local optimization of the environmental parameters to considering the entire supply chain production process, consumption, customer service, and alternative channel structures (Linton et al. 2007; Xue et al. 2014).

There are numerous studies, especially during the 1990s, which deal with the alternative aspect of green channels. These channels grew from the consumers' desire to benefit from all the positive elements of the 'green' concept, such as eco-friendly products, health and safety issues, and environmental protection. The integration of the environmental factor in distribution channels is today a fact worldwide, whose positive impact not only influences consumers but has great value for businesses and producers as well. According to Boks and Stevens (2007) there are three different 'green' categories, namely science, government, and customers that are based on the different perceptions that each of the involved parties have of the environment. A number of studies have investigated which factors compel companies to redesign their supply chain in a more environmentally friendly and responsible way (Bansal and Roth 2000; Hoffman 2001). Interestingly, this showed the crucial factor that has played an essential role in leading companies to adopt a green policy was legislation on environmental issues (Paulraj 2009; Rao and Holt 2005). Yet another factor that motivates companies to be environmentally-friendly is the intense competition that exists in the current business world (Morrow and Rondinelli 2002). However, it is not only economic factors that instigate becoming 'green' but according to Bansal and Roth (2000) moral factors also influence this transformation.

As regards agricultural distribution, the contraction of the agro-food chain with smaller, local distribution channels and the adoption of direct sales by the farm, promotes green logistics, thus contributing to environmental protection (Gilg and Battershill 2000). Shortening the agricultural distribution chain so that the products reach the customer directly not only reduces transportation distances that places a heavy burden on the environment but also cuts external financial costs for producers and consumers.

In sum, the relatively broad category of 'green' as an alternative form of agricultural distribution channel is chiefly characterised by environmental factors, which are based on both economic and moral dimensions. This has given rise to green or ethical markets that provide the consumer with a variety of commodities, including produce from sustainable farming, such as free-range poultry and eggs, organic dairy products, etc. (Lekakis 2014). There has been an increasing need for society to find sources of sustainable food systems. This has meant returning to basics, such as organic farming, eco-friendly packaging, and environmentally-conscious distribution of products. Thus, the category 'green' as a form of alternative distribution channel entwines all the others that follow, in the sense that 'green' attitudes and behaviors in combination with environmental factors have now been widely and deeply established in society.

3.2 'E-commerce' Channel

As a distribution channel 'e-commerce' is alternative in that it enables consumers to shop online at anytime and from anywhere in the world, often at more economical prices. The concept of 'e-commerce' was established to meet the needs of some consumers, who for a variety of reasons were not able to shop at physical stores. This gradually evolved to encompass ever-wider sectors of society, largely due to the benefits of online shopping, which are convenience, time-saving, fewer 'luring' traps, variety and better prices. 'E-commerce' as a distribution channel has grown to encompass the agrifood sector.

The technological changes that are occurring have had the result of making agrifood systems much more complex (Lowe et al. 2008). Thus, a growing body of researchers have been involved in developing knowledge-based information systems that help managers in the on-line decision-making process generally (Matsatsinis and Siskos 1999; Özbayrak and Bell 2003), while Wen (2007) has built an "intelligent knowledge-based e-commerce system" specifically for the sale of agricultural products.

On the other hand, recent research has shown that customers' online shopping behavior has a large impact on the conduct of e-commerce systems. Mitrevski and Hristoski (2014) identify consumers' behavioural factors as the most essential subsystem of e-commerce. An essential condition is for it to be compatible with the lifestyle, experiences and purchasing habits of Internet users (O'Cass and Fenech 2003). To overcome the obstacles in the development of the Internet and e-commerce and for it to grow even more, the consumer must have added value through online shopping (Anckar et al. 2002). E-commerce is a real benefit to consumers as an alternative shopping channel when the following elements apply: (a) competitive prices (b) more convenient marketing procedures, (c) superior service provision to consumers, and (d) wide variety and availability of specialised goods (Anckar et al. 2002). Another obvious reason that has established this as an alternative type of purchasing—agricultural products included—is the fact that many consumers no longer seem to have the time to physically go to outlets, or prefer to use their available time for other activities apart from shopping.

Consumer concerns related to online shopping are well-known. Research has shown that despite the risks being more psychological than real, consumers give security—of their credit card information, personal data, correct processing of their order (Lau et al. 2010), and hidden costs of taxes and tariffs (Lim 2003)—as the main reason for their reluctance to make online purchases (Bourdeau et al. 2002). In addition, the legal framework for the protection of the online buyer is often unclear and at present incomplete (Whysall 2000), Internet fraud widely publicized in the media, the lack of consumer awareness regarding encryption mechanisms, and the absence of a standard method for electronic payment of goods purchased (Vijayasarathy and Jones 2000) are further serious limitations to e-commerce.

In contrast, customers tend to have more trust in well-known businesses or those with a good reputation, and they feel more secure in dealing with web-based companies that provide a telephone contact number and a physical address (Linton et al. 2007; Lim 2003). In the literature, trust is regarded as the driving force in the buyer-seller relationship, and is described as a prerequisite for achieving maximum performance and competitive advantage (Abatekassa 2011). According to Van de Poel and Leunis (1999) three ways to reduce the risk of the Internet market are: (a) selling at reduced prices, (b) offering well-known/famous brands, and (c) providing a money-back-guarantee.

In relation to agriculture, it must be stated that this type of alternative distribution channel has the major drawback that products cannot be touched, felt or smelt. In addition, as there is no human contact, there cannot be one-to-one interaction, making the entire process not only impersonal but also of limited scope due to the perishable nature of the products being provided. Thus, the online purchase of fresh fruit and vegetables must occur on or close to the same day and locally or at most regionally. On the whole, despite the fact that in order to shop online one must not only possess the necessary technology but also be e-literate. As a form of alternative distribution, e-commerce has very quickly become an accepted means of product distribution for both providers and consumers (Wen 2007).

3.3 'Local Food Networks' Channel

'Local Food Networks' is an alternative distribution channel because it offers consumers fresh produce that is grown in the community or region, as opposed to imported or freezer goods. Although rural areas have had informal agrifood networks operating in their small communities for a long time, this alternative channel has spread to urban areas as a result of a change in consumer mentality. The qualities of 'Local Food Networks' are many. Apart from the evident economic benefits to producers, consumers and communities, there are the less obvious non-economic benefits, such as educational knowledge of farming, recreational activities (community gardening), and increased social interaction (between different racial, economic, age groups in the community) that the members gain. In addition, there are the health benefits to consumers and communities, as well as the environmental benefits, from the reduction in carbon dioxide emissions, and sustainable energy use.

Local and/or regional food networks focus on sustainable food production, distribution and consumption through the efforts of collaboration in order to establish local, self-reliant economies (Katchova and Woods 2013). In contrast to globalization, localization places prominence on economic, environmental and social health enhancement of specific localities or regions (ibid). Local/regional emphasis not only benefits farmers but also the community, as well as the consumer. In a recent study Cleveland et al. (2014) found that localization has become a favored strategy that gives alternative solutions, which lends support to our having placed this type of distribution channel in the sub-category of "new trend".

Farmers' markets are perhaps the oldest type of buyer—seller relationship. They are physical markets where producers of agricultural products set up stalls and sell directly to the customer. These markets around the world are usually held outdoors and are characterized by a local culture and economy (Conner 2010). The advantage of this approach for the farmer is that the final retail price is guaranteed for both parties, since the farmer/producer comes into direct contact with the consumer (Govindasamy 2002) cutting out any intermediaries. Of particular interest are Brown's findings (2002) that small holdings are three times more likely to use direct sales to consumers rather than intermediaries. In addition, farmers' markets as a distribution channel deal with the alternative food system of localization which increases sustainability (Cleveland et al. 2014). On the other hand, it has been suggested that an optimal policy to help farmers/producers either become or remain economically viable would be for them to have access to government grants (Aksen et al. 2009), which would provide incentives for genuine sustainability in their business ventures (Mitra and Webster 2008). The provision of government funds is a much more preferable option than attempts to intervene in the rhythm of harvesting and profit levels through the implementation of legislation (Aksen et al. 2009).

The recent economic recession has seen a new surge in consumer preference for farmers' markets. An interesting case is that of Greece were it has become an alternative distribution channel arising in response to the on-going economic crisis. More specifically, in 2012 a volunteer grass-roots action group in Pieria (in the Region of Central Macedonia, Northern Greece) established a self-organised co-operative—the so-called Potato Movement—which allowed producers to sell directly to the public bypassing the intermediaries. The movement has since become nationwide and its main aim was for agricultural producers to sell directly to the public in an attempt to eliminate the unfair profiting of wholesalers that was at the detriment of both farmers and consumers, as well as being a way to provide some relief to rising living costs in the country's worsening economic situation (Henley 2012).

On the other hand, food co-operatives (co-ops) as a type of organizational association go back centuries, agricultural marketing co-ops, however, came later (Kimberly 1980). Cooperatives are a particular business model, owned by the members who operate them collaboratively (Lund 2013), they play an important role in sustainable food production and distribution (Katchova and Woods 2013). They are involved primarily in sourcing and marketing local or regional products.

Interestingly, there seems to be an increasing trend, where consumers prefer food co-ops as a retailing mechanism (ibid).

Although there have been some concerns as to the economic effectiveness of local food networks as supply chains (Lusk and Norwood 2011), overall they have been shown to apply successful competitive business strategies in linking local farmers/producers with consumers. They have, thus, been able to form "highly differentiated markets and local supply chains" (Katchova and Woods 2013: 241), making them an operative alternative distribution channel. In sum, local food networks that incorporate farmers' markets and food cooperatives are a newer form of alternative agricultural distribution channel primarily because of their localized nature. As an alternative food system it increases sustainability, as well as promotes the direct farmer/producer—consumer relationship that cuts out the middleman. Thus, 'Local Food Networks' is a form of alternative distribution channel that incorporates Farmers' Markets and Food Cooperatives.

3.4 'Local Food Networks' Channel

'The community' comprises Fair Trade, which, on the one hand, assists producers in developing nations to gain more equitable trading conditions, and on the other, encourages sustainability. It constitutes an alternative distribution channel because consumers make a conscious decision to support these emerging economies instead of established ones. The recent developments in the field of human geography incorporate a convergence between the 'economic' and the 'cultural', which deals with social aspects (Jackson 2002). Social and political progressive causes, such as the fair trade movement, have sprung from the new field of the politics of consumption (Micheletti and Stolle 2012), which is associated with people's rights to an acceptable standard of living, quality of life, equal participation opportunities, as well as the notion of ecological sustainability. The historical roots of the fair trade crusade along with effective promotional and publicity campaigns have made it a vanguard of social and political global issues as it is closely associated with the notion of sustainable futures and sustainable citizenship (Micheletti and Stolle 2012).

For almost a quarter of a century, the fair trade movement has developed from a loose, informal group of social activists and producers into an important organized international network that has established several recognized certification programs. More and more researchers are dealing with the critical issue of certification and ethical consumption of food, focusing to a large extent on food sources that derive from sustainable forest products, organic farming and Fair Trade (McCarthy 2006; Renard 2003, 2005; Levi and Linton 2003; Getz and Shreck 2006; Taylor 2005; Mutersbaugh 2002, 2005; Klooster 2005; Herman 2010; Barham et al. 2011). They have tended to focus on the certification of primary food products, such as, coffee, tea, cocoa, sugar, honey, bananas and other fresh and dried fruit (Mutersbaugh 2005; Taylor 2005; Renard 2005; Getz and Shreck 2006; Eden 2009). However, even though there are works that focus on the ethical concerns of production and trading methods for farmers and artisans, there is no clear orientation on the function of certification (Littrell and Dickson 1997, 1998).

It is an undeniable fact that there has been a remarkable increase in the sales of fair trade products in mainstream markets, particularly for foodstuffs, the most prevalent being tea, coffee and bananas. To give an example of the sheer size of this increase, according to the Fair Trade Foundation (2013) the yearly retail sales of fair trade products are now estimated as being over one billion pounds in Great Britain alone. Overall, when compared to the other distribution means, fair trade clearly has other social, economic and political dimensions. More specifically, human rights and social justice are deeply engrained in the principles of fair trade, the main objectives of which are to help farmers, workers and artisans in developing countries not only earn a respectable living but also express a sense of happiness, as well as focusing on ecological sustainability (Becchetti et al. 2011; Biggs and Lewis 2009; Becchetti 2010). These are the major characteristics that constitute fair trade as an alternative channel, particularly for agricultural products, which are slowly but steadily gaining momentum all over the world and not just the most developed nations. Thus, 'the community' is a form of alternative distribution channel.

4 Summary of Results and Further Study Considerations

An overall concluding remark is that the field of alternative distribution and marketing of products is quite broad including all those forms that are different to mainstream. For example, in the case of alternative tourism, anything other than mass tourism should be graced with the alternative label (de Kadt 1990). The same applies in the case of alternative distribution channels of agricultural products. Economic forces and conditions, both at a global and local level, have influenced consumer attitudes and behavior, which has brought about changes in their choice of distribution channels. There has been a growing tendency towards alternative channels over mainstream ones. The result has been alternative product distribution channels being separated into the 'established" and "new trend" categories. The four forms which constitute the key concepts are: 'green' that comprises environmental aspects of sustainability (see Sect. 3.1), 'e-commerce' that involves on-line trade (see Sect. 3.2), 'local food networks' that focus on local, self-reliant economies based on sustainable food production and consumption (see Sect. 3.3), and 'the community' which incorporates fair trade stemming from the politics of consumption (see Sect. 3.4).

Alternative agricultural product distribution channels have been found to play a very important role in the spreading of social and economic advantages at each stage of the production process, since they facilitate the direct relationship between producers and consumers. In fact, it occurred that the alternative distribution channels listed in the present paper are networks for opportunity at a socio-economic and environmental level. As Jackson argued, production and consumption, the local and the global, or culture and economy have a mutual structure and should not be viewed as opposites (Jackson 2002).

On the whole, the growing trend for customers to turn to alternative forms of distribution channels especially for agricultural products stems from the following: a preference for local produce that is fresher and more economical ('local food networks' distribution channel); a conscious contribution to a fairer and more sustainable society ('the community' distribution channel); a faster, easier and more digital means of purchasing ('e-commerce' distribution channel); and a growing awareness of the need to be more environmentally active ('green' distribution channel).

The contribution of the present paper, which is a review of research into the field of distribution channels, is that it sheds light on the definition of 'alternative' channels. In the hitherto literature, there is no clear definition regarding this term. Thus, the objective of this paper, which was to investigate what constitutes 'alternative' distribution channels in the agri-food system has contributed to this end.

This creates an incentive for future research to also examine consumer attitudes and behavior related to these four distribution forms, which will further contribute to the field of 'alternative' distribution channels. In addition, of particular interest presents the hypothesis for further study that these four forms of alternative distribution channels have an impact on the creation of new employment opportunities.

5 Conclusion

Overall, the agricultural product distribution channels that incorporate 'green', 'e-commerce', 'local food networks' and 'the community' have the common trait of being alternative. Although each distribution means can stand alone, there can and often there is an overlap in certain characteristics. In other words, in terms of the 'green' issue, farmers' markets and food cooperatives can sell green products, and green produce or organic fair trade commodities can be purchased online. Short supply chains, usually by reducing intermediaries, are another factor shared by farmers' markets and e-commerce, for instance. In addition, the issues of localization and sustainability are connected with all.

On the whole, the key factor that seems to permeate all four forms in most of the articles reviewed is the concept of sustainability. Sustainable management and development of agricultural product distribution channels take into consideration the environmental, societal, and economic dimensions. It cannot be denied that they play a very important role in promoting social, environmental and economic benefits for farmers/producers, consumers and the community as a whole, and for this reason are becoming an increasingly preferred product distribution mechanism.

The added value of this paper is that we were able to contribute to a more concise explanation as to what constitutes alternative distribution channels of agricultural products. As derived from this study, for a distribution channel to be considered alternative, it must have one of the four forms, which differentiates it from mainstream channels. Additionally, from the study's overall findings, a set of guidelines was established for managers and investors to assist in implementing an alternative distribution channel. The S.M.A.R.T. checklist stands for: Sustainability, it must be eco-friendly; Maximization of profits, which leads to further investments; Accessibility of product to consumers; Responsibility for high quality product to consumers; Trust, provide optimal trust in consumer-manager-producer relationship.

Consumers are becoming more aware of the food system and its association to the natural environment, local economies, and sustainable futures. It appears then that there is a new trend in their actively seeking out 'alternatives' to mainstream or conventional means. Therefore, there is a growing tendency for customers to take a more active role even in their choice of distribution channels, especially when agricultural products are involved.

In sum, it is of the utmost importance in today's market to define what constitutes alternative product distribution channels in order to meet the needs of both the producer and the consumer. The findings of the present study present data can be adopted to draw up producer quality standards. The implications of the present study on alternative channels in the agrifood sector are that it provides valuable information to all players in the field, which can be applied to advantage. Producers can become acquainted first-hand with customer attitudes and needs; knowledge which allows them to make innovative, profitable choices that are both timely and cost effective. Consumer organizations and interested individual consumers can learn about the existing options which can help in informed decision-making regarding purchasing trends. Additionally, the study findings are associated to food supply, including supply chain and logistics and food sustainability, as well as fostering consumer choice concerning alternative distribution channels of agricultural products. Lastly, it could even propose strategic directions to policy makers by providing them with current trends that will assist them to formulate and implement the most appropriate government and business schemes in terms of legislation, organization and regulation.

References

Abatekassa, G., & Peterson, C. (2011). Market access for local food through the conventional food supply chain. *International Food and Agribusiness Management Review, 14,* 63–82.

Achrol, R. (1997). Changes in the theory of interorganizational relations in marketing: Toward a network paradigm. *Journal of the Academy of Marketing Science, 25,* 56–71.

Aksen, D., Aras, N., & Karaarslan, A. G. (2009). Design and analysis of government subsidized collection systems for incentive-dependent returns. *International Journal of Production Economics, 119,* 308–327.

Anckar, B., Walden, P., & Jelassi, T. (2002). Creating customer value in online grocery shopping. *International Journal of Retail & Distribution Management, 30,* 211–220.

Ashby, A., Leat, M., & Melanie Hudson-Smith, M. (2012). Making connections: A review of supply chain management and sustainability literature. *Supply Chain Management: An International Journal, 17*, 497–516.

Bansal, P., & Roth, K. (2000). Why companies go green: A model of ecological responsiveness. *Academy of Management Journal, 43*, 717–736.

Bao, L., Huang, Y., Ma, Z., Zhang, J., & Lv, Q. (2012). On the supply chain management supported by e-commerce service platform for agreement based circulation of fruits and vegetables. *Physics Procedia, 33*, 1957–1963.

Barham, B. L., Callenes, M., Gitter, S., Lewis, J., & Weber, J. (2011). Fair trade/organic coffee, rural livelihoods, and the "Agrarian Question": Southern Mexican coffee families in transition. *World Development, 39*, 134–145.

Baxter, R. (2012). How can business buyers attract sellers' resources? Empirical evidence for preferred customer treatment from suppliers. *Industrial Marketing Management, 41*, 1249–1258.

Becchetti, L., Castriota, S., & Solferino, N. (2011). Development projects and life satisfaction: An impact study on fair trade handicraft producers. *Journal of Happiness Studies, 12*, 115–138.

Becchetti, L., & Michetti, M. (2010). When fair trade generates social capital by creating room for manoeuvre for pro-poor policies. *African Journal of Business Management, 4*, 2903–2914.

Biggs, S., & Lewis, D. (2009). Fair trade and organizational innovation in nepal: Lessons from 25 years of growth of the association of craft producers (ACP). *European Journal of Development Research, 21*, 377–396.

Blonska, A. (2010). *To buy or not to buy: Empirical studies on buyer-supplier collaboration.* Maastricht: Universitaire Pers.

Boks, C., & Stevels, A. (2007). Essential perspectives for design for environment. Experiences from the electronics industry. *International Journal of Production Research, 45*, 4021–4039.

BORDEN, N. H. 1965. The Concept of the Marketing Mix.

Borghi, A., Gallo, M., Strazza, C., & del Borghi, M. (2014). An evaluation of environmental sustainability in the food industry through life cycle Assessment: The case study of tomato products supply chain. *Journal of Cleaner Production, 78*, 121–130.

Bourdeau, L., Chebat, J.-C., & Couturier, C. (2002). Internet consumer value of university students: E-mail-vs.-Web users. *Journal of Retailing and Consumer Services, 9*, 61–69.

Brown, A. (2002). Farmer's market research 1940–2000: An inventory and review. *American Journal of Alternative Agriculture, 17*, 167–176.

Burgess, K., Singh, P. J., & Koroglu, R. (2006). Supply chain management: a structured literature review and implications for future research. *International Journal of Operations & Production Management, 26*, 703–729.

Burt, L., & Wolfley, B. (2009). Farmer-to-consumer marketing: The series, PNW–206, Oregon State University. Oregon State University, Washington State University, University of Idaho.

Cantor, D. E., Blackhurst, J. V., & Cortes, J. D. (2014). The clock is ticking: The role of uncertainty, regulatory focus, and level of risk on supply chain disruption decision making behavior. *Transportation Research Part E: Logistics and Transportation Review, 72*, 159–172.

Chapman, R., L., Claudine, S., & Jay, K. (2003). Innovation in logistic services and the new business model. *International Journal of Physical Distribution & Logistics Management, 33*, 630–650.

Cheng, L.-C., & Grimm, C. M. (2006). The application of empirical strategic management research to supply chain management. *Journal of Business Logistics, 27*, 1–55,

Chinnici, G., D'Amico, M., & Pecorino, B. (2002). A multivariate statistical analysis on the consumers of organic products. *British Food Journal, 104*, 187–199.

Cleveland, A., Nora, M., Alexander, C., Mazaroli, D., & Hinson, K. (2014). Local food hubs for alternative food systems: A case study from Santa Barbara County, California. *Journal of Rural Studies, 35*, 26–36.

Conner, D., Colasanti, K., Ross, B., & Smalley, B. (2010). Locally grown foods and farmers markets: Consumer attitudes and behaviors. *Sustainability, 2*, 742–756.

Cordano, M. (1993). Making the natural connection: Justifying investment in environmental innovation. In *Proceedings of the International Association for Business and Society* (pp. 530–537).

de Kadt, E. (1990). *Making the alternative sustainable: Lessons from development for tourism* (pp. 1–21). Discussion Paper 272, Institute of Development Studies, Brighton.

Dekker, H. C., Sakaguchi, J., & Kawai, T. (2013). Beyond the contract: Managing risk in supply chain relations. *Management Accounting Research, 24,* 122–139.

Denyer, D., Tranfield, D., Bryman, A., & Buchanan, D. (2009). *Producing a systematic review.* London: Sage Publications.

Dimara, E., & Skuras, D. (2005). Consumer demand for informative labeling of quality food and drink products: A European Union case study. *Journal of Consumer Marketing, 22,* 90–100.

Eden, S. (2009). The work of environmental governance networks: Traceability, credibility and certification by the Forest Stewardship Council. *Geoforum, 40,* 383–394.

Elith, J. A., & Leathwick, J. (2009). Species distribution models: Ecological explanation and prediction across space and time. *Annual Review of Ecology, Evolution, and Systematics, 40,* 677–697.

Ellis, S. C., Henke Jr., J. W., & Kull, T. J. (2012). The effect of buyer behaviors on preferred customer status and access to supplier technological innovation: An empirical study of supplier perceptions. *Industrial Marketing Management, 41,* 1259–1269.

Essig, M., & Amann, M. (2009). Supplier satisfaction: Conceptual basics and explorative findings. *Journal of Purchasing and Supply Management, 15,* 103–113.

Falguera, V., Aliguer, N., & Falguera, M. (2012). An integrated approach to current trends in food consumption: Moving toward functional and organic products? *Food Control, 26,* 274–281.

Fink, A. (2009). *Conducting research literature reviews: From the internet to paper.* CA: Thousand Oaks.

French, S., Maule, J., & Papamichail, N. (2009). *Decision behaviour, analysis and support.* Cambridge: Cambridge University Press.

Getz, C., & Shreck, A. (2006). What organic and fair trade labels do not tell us: Towards a place-based understanding of certification. *International Journal of Consumer Studies, 30,* 490–501.

Gilg, A. W., & Battershill, M. (2000). To what extent can direct selling of farm produce offer a more environmentally friendly type of farming? Some evidence from France. *Journal of Environmental Management, 60,* 195–214.

Govindasamy, R., Italia J., & Adeleja, A. (2002). Farmers markets: Consumer trends, preferences and characteristics. *Journal of Extension, 40.*

Guertler, B., & Spinler, S. (2015). When does operational risk cause supply chain enterprises to tip? A simulation of intra-organizational dynamics. Omega.

Heberling, M. T., Templeton, J. J., & Wu, S. (2012). Green net regional product for the San Luis Basin, Colorado: An economic measure of regional sustainability. *Journal of Environmental Management, 111,* 287–297.

Heckmann, I., Comes, T., & Nickel, S. (2015). A critical review on supply chain risk—Definition, measure and modeling. *Omega, 52,* 119–132.

Henley, J. (2012). Greece on the breadline: 'Potato movement' links shoppers and farmers. *The Guardian.*

Herman, A. (2010). Connecting the complex lived worlds of fairtrade. *Journal of Environmental Policy & Planning, 12,* 405–422.

Hoffman, A. J. (2001). Linking organizational and field-level analyses: The diffusion of corporate environmental practice. *Organization & Environment, 14,* 133–156.

Hughes, A. (2005). Geographies of exchange and circulation: Alternative trading spaces. *Progress in Human Geography, 29,* 496–504.

Jackson, P. (2002). Commercial cultures: Transcending the cultural and the economic. *Progress in Human Geography, 26,* 3–18.

Jain, J., Dangayach, G., Agarwal, G., & Banerjee, S. (2010). Supply chain management: literature review and some issues. *Journal of Studies on Manufacturing, 1,* 11–25.

Jamaluddin., N. (2013). Adoption of E-commerce practices among the Indian farmers, a survey of Trichy district in the state of Tamilnadu, India. *Procedia Economics and Finance, 7,* 140–149.

Katchova, A., & Woods, T. (2013). Local foods and food cooperatives: Ethics, economics and competition issues. In: J. H. S. James (Ed.), *The ethics and economics of agrifood competition.* Netherlands: Springer.

Kimberly A., Z., Cropp, R. (1980). Cooperatives: Principles and practices in the 21st century. Wisconsin: University of Wiscosin Extension—Madison.

Klooster, D. (2005). Environmental certification of forests: The evolution of environmental governance in a commodity network. *Journal of Rural Studies, 21,* 403–417.

Komatsu, Y., Tsunekawa, A., Ju, H. (2005). Evaluation of agricultural sustainability based on human carrying capacity in drylands—A case study in rural villages in Inner Mongolia, China. *Agriculture, Ecosystems and Environment, 108,* 29–43.

Kovacs, G., Spens, K., Mortensen, M., Freytag, P., & Stentoft, J. (2008). Attractiveness in supply chains: A process and matureness perspective. *International Journal of Physical Distribution & Logistics Management, 38,* 799–815.

Ladhari, R., & Tchetgna, N. M. (2015). The influence of personal values on Fair Trade consumption. *Journal of Cleaner Production, 87,* 469–477.

Lau, H., Jiang, Z., & Wang, D. (2010). A credibility-based fuzzy location model with Hurwicz criteria for the design of distribution systems in B2C e-commerce. *Computers & Industrial Engineering, 59,* 873–886.

Lauterborn, B. (1990). New marketing litany: Four Ps Passé: C-words take over. *Advertising Age, 61,* 26.

Lawson, B., Cousins, P., Squire, B., Burgess, K., Singh, P. J. & Koroglu, R. (2006). Supply chain management: A structured literature review and implications for future research. *International Journal of Operations & Production Management, 26,* 703–729.

Lekakis, E. J. (2014). ICTs and ethical consumption: The political and market futures of fair trade. *Futures, 62,* Part B, 164–172.

Levi, M., & Linton, A. (2003). Fair trade: A cup at a time? *Politics & Society, 31,* 407–432.

Leyshon, A. (2005). Introduction: Diverse economies. *Antipode, 37,* 856–862.

Leyshon, A., Lee, R., & Williams, C. C. (2003). *Alternative economic spaces.* London: Sage.

Lim, N. (2003). Consumers' perceived risk: Sources versus consequences. *Electronic Commerce Research and Applications, 2,* 216–228.

Linton, J. D., Klassen, R., & Jayaraman, V. (2007). Sustainable supply chains: An introduction. *Journal of Operations Management, 25,* 1075–1082.

Littrell, M., & Dickson, M. (1997). Alternative trading organizations: Shifting paradigm in a culture of social responsibility. *Human Organization, 56,* 344–352.

Littrell, M. A., & Dickson, M. A. (1998). Fair trade performance in a competitive market. *Clothing and Textiles Research Journal, 16,* 176–189.

Lowe, P., Phillipson, J., & Lee, R. P. (2008). Socio-technical innovation for sustainable food chains: Roles for social science. *Trends in Food Science & Technology, 19,* 226–233.

Lund, M. (2013). *Cooperative equity and ownership: An introduction.*

Lusk, J. L., & Norwood, F. B. (2011). The Locavore's dilemma: Why pineapples shouldn't be grown in North Dakota. *Library of economics and liberty.* Retrieved September 1, 2011 from http://www.econlib.org/library/Columns/y2011/LuskNorwoodlocavore.html.

Matsatsinis, N. F., & Siskos, Y. (1999). MARKEX: An intelligent decision support system for product development decisions. *European Journal of Operational Research, 113,* 336–354.

Mattas, K., Tsakiridou, E, (2010). Shedding fresh light on food industry's role. The recession's aftermath. *Trends in food Science and Technology,* 212–216.

Mauleón, J.-R. (2003). Contribution of short food chains to rural development in the Basque Country. In *XX Congress of the European Society for Rural Sociology, Working Group 1.2. Food consumption and farming., University of the Basque Country, Department of Sociology* (Vol. 2).

McCarthy, J. (2006). Rural geography: Alternative rural economies—The search for alterity in forests, fisheries, food, and fair trade rural geography: Alternative rural economies—the search for alterity in forests, fisheries, food, and fair trade. *Progress in Human Geography, 30,* 803–811.

Meier, M. S., Stoessel, F., Jungbluth, N., Juraske, R., Schader, C., & Stolze, M. (2015). Environmental impacts of organic and conventional agricultural products—Are the differences captured by life cycle assessment? *Journal of Environmental Management, 149,* 193–208.

Micheletti, M., & Stolle, D. (2012). Sustainable citizenship and the new politics of consumption. *The ANNALS of the American Academy of Political and Social Science, 644,* 88–120.

Mitra, S., & Webster, S. (2008). Competition in remanufacturing and the effects of government subsidies. *International Journal of Production Economics, 111,* 287–298.

Mitrevski, P. J., & Hristoski, I. S. (2014). Behavioral-based performability modeling and evaluation of e-commerce systems. *Electronic Commerce Research and Applications, 13,* 320–340.

Morrow, D., & Rondinelli, D. (2002). Adopting corporate environmental management systems: Motivations and results of ISO 14001 and EMAS certification. *European Management Journal, 20,* 159–171.

Mutersbaugh, T. (2002). The number is the beast: A political economy of organic-coffee certification and producer unionism. *Environment and Planning A, 34,* 1165–1184.

Mutersbaugh, T. (2005). Just-in-space: Certified rural products, labor of quality, and regulatory spaces. *Journal of Rural Studies, 21,* 389–402.

O'Cass, A., & Fenech, T. (2003). Web retailing adoption: Exploring the nature of internet users Web retailing behaviour. *Journal of Retailing and Consumer Services, 10,* 81–94.

Özbayrak, M., & Bell, R. (2003). A knowledge-based decision support system for the management of parts and tools in FMS. *Decision Support Systems, 35,* 487–515.

Panayides, P. M., & Venus Lun, Y. H. (2009). The impact of trust on innovativeness and supply chain performance. *International Journal of Production Economics, 122,* 35–46.

Paulraj, A. (2009). Environmental motivations: A classification scheme and its impact on environmental strategies and practices. *Business Strategy and the Environment, 18,* 453–468.

Pereira, J. V. (2009). The new supply chain's frontier: Information management. *International Journal of Information Management, 29,* 372–379.

Porter, M. E. (1985). *Competitive advantage: Creating and sustaining superior performance.* New York.

Rao, P., & Holt, D. (2005). Do green supply chains lead to competitiveness and economic performance? *International Journal of Operations & Production Management, 25,* 898–916.

Renard, M.-C. (2003). Fair trade: Quality, market and conventions. *Journal of Rural Studies, 19,* 87–96.

Renard, M.-C. (2005). Quality certification, regulation and power in fair trade. *Journal of Rural Studies, 21,* 419–431.

Roy, S., Sivakumar, K., & Wilkinson, I. (2004). Innovation generation in supply chain relationships: A conceptual model and research propositions. *Journal of the Academy of Marketing Science, 32,* 61–79.

Schiele, H. (2008). Location, location: The geography of industry clusters. *Journal of Business Strategy, 29,* 29–36.

Schreiber, R., Crooks, D., & Stern, P. N., et al. (1997). Qualitative meta—Synthesis; issues and techniques. In J. M. Morse (Ed.), *Completing a qualitative project; details and dialogue.*

Sheffi, Y. (2001). Supply chain management under the threat of international terrorism. *International Journal of Logistics Management, 12.*

Sheng, J., Shen, L., Qiao, Y., Yu, M., & Fan, B. (2009). Market trends and accreditation systems for organic food in China. *Trends in Food Science & Technology, 20,* 396–401.

Shu., G., Tian-Zhi., R., & Mao-Hua., W. (2007). Technology and infrastructure considerations for e-commerce in chinese agriculture. *Agricultural Sciences in China, 6,* 1–10.

Solomon, M. (2009). *Consumer behavior, buying, having and being.* New Jersey, USA: Pearson Education Inc.

Soosay, C. A., Hyland, P. W., & Ferrer, M. (2008). Supply chain collaboration: Capabilities for continuous innovation. *Supply Chain Management: An International Journal, 13,* 160–169.

Spiller, F., Zuhlsdorf, A., Mellin, M. (2008). Customer satisfaction in farmer-to consumer direct marketing. *International Food and Agribusiness Management Association, 11*.

Strauss, A., Corbin, J. (1998). *Basics of qualitative research*. SAGE Publications.

Svensson, G. (2000). A conceptual framework for the analysis of vulnerability in supply chains. *International Journal of Physical Distribution & Logistics Management, 30*, 731–750.

Sydorovych, O., & Wossink, A. (2008). The meaning of agricultural sustainability: Evidence from a conjoint choice survey. *Agricultural Systems, 98*, 10–20.

Taylor, P. L. (2005). In the market but not of it: Fair trade coffee and forest stewardship council certification as market-based social change. *World Development, 33*, 129–147.

Tranfield, D., Denyer, D., & Smart, P. (2003). Towards a methodology for developing evidence-informed management knowledge by means of systematic review. *British Journal of Management, 14*, 207–222.

Tsang, E. W. K. (2000). Transaction cost and resource-based explanations of joint ventures: A comparison and synthesis. *Organization Studies, 21*, 215–242.

van den Poel, D., & Leunis, J. (1999). Consumer acceptance of the internet as a channel of distribution. *Journal of Business Research, 45*, 249–256.

Veatch, B., & Maren, E. (2007). In S. Raman (Ed.), *Agricultural sustainability: Principles, processes, and prospects*. Food Products Press, Haworth Press Inc., 10 Alice Street, Binghamton, NY 13904–1580, USA (2006); *Industrial Crops and Products, 26*, 105–106.

Vijayasarathy, L. R., & Jones, J. M. (2000). Print and internet catalog shopping: Assessing attitudes and intentions. *Internet Research, 10*, 191–202.

Watts, M. (2003). Development and governmentality. *Singapore Journal of Tropical Geography, 24*, 6–34.

Wen, W. (2007). A knowledge-based intelligent electronic commerce system for selling agricultural products. *Computers and Electronics in Agriculture, 57*, 33–46.

Whatmore, S., Stassart, P., & Renting, H. (2003). What's alternative about alternative food networks. *Environment and Planning, 35*, 389–391.

Whysall, P. (2000). Retailing and the internet: A review of ethical issues. *International Journal of Retail & Distribution Management, 28*, 481–489.

Wisner, J. D. (2003). A structural equation model of supply chain management strategies and firm performance. *Journal of Business Logistics, 24*, 1–26.

Xue, W., Caliskan Demirag, O. & Niu, B. (2014). Supply chain performance and consumer surplus under alternative structures of channel dominance. *European Journal of Operational Research, 239*, 130–145.

Yu, W., Chavez, R., Feng, M., & Wiengarten, F. (2014). Integrated green supply chain management and operational performance. *Supply Chain Management: An International Journal, 19*, 683–696.

Zsidisin, G. A., Panelli, A. & Upton, R. (2000). Purchasing organization involvement in risk assessments, contingency plans, and risk management: An exploratory study. *Supply Chain Management: An International Journal, 5*, 187–198.

Kallirroi Nikolaou Kallirroi Nikolaou is Ph.D. candidate, in the Department of Agricultural Economics, from Aristotle University of Thessaloniki. Her education includes a BSc in Agriculture from the Aristotle University of Thessaloniki (2011), an MSc in Agricultural Economics from the Aristotle University of Thessaloniki (2012). Her research interests focus on alternative distribution channels of agricultural products.

Foivos Anastasiadis Foivos Anastasiadis is a Postdoctoral Researcher at Aristotle University of Thessaloniki (A.U.Th.). He was specifically recruited as an expert on sustainable agrifood supply chains for the EU funded FP7 project GREEN-AgriChains. He holds an MSc from Wagenigen University, The Netherlands (2004) and a Ph.D. from Imperial College, London (2010). Prior to joining A.U.Th, Dr. Anastasiadis was employed: as an adjunct lecturer at the University of

Macedonia, Greece; as an academic associate at the International Hellenic University, Greece; and as a consultant in the market research sector and in the supply chain/procurement processes business.

Efthimia Tsakiridou Efthimia Tsakiridou is an Associate Professor at the Aristotle University of Thessaloniki, Greece. She has twelve years of research experience in the field of food marketing and supply chain management. She is a specialist in the field of agro-food marketing, supply chain management and consumer behavior and worked as a participant in several European and national research projects either as a participant or as a leading researcher. Dr. Tsakiridou has published in several refereed journals (Applied Economics, Journal of Food Products Marketing, Journal of International Food and Agribusiness Marketing, International Journal of Retail and Distribution Management, British Food Journal, Food Economics, International Journal of Economic Research).

Konstadinos Mattas Konstadinos Mattas is a Professor of Agricultural Policy. His education includes a B.Sc. in Agriculture from the Aristotle University of Thessaloniki (1972), a B.Sc. in Economics from the University of Macedonia (1976), an MSc in Agricultural Economics from the University of Kentucky (1982) and a Ph.D. in Agricultural Economics from the University of Kentucky (1984). He has published in more than 100 international refereed journals, in collective volumes and proceedings. He has contributed to international and Greek conferences and served as a referee to several international journals.

Opportunities of Price Risk Limitation in Horticultural Sector in Poland

Lukasz Zaremba

Abstract This paper reviews the possibility of price risk limitation, which is conducive to farmers' income stabilization. The attempts to control price risk in both the agricultural and horticultural sectors started a long time ago. The price is the most important factor bounding supply and demand on the market. Prices of agri-food products are volatile and derive from many factors, which are unable to be predicted in practice. The influence of many variables is, furthermore, also volatile and emerges in varying intensity. Price risk in the farmers' case results from the fact that the production decisions are made mostly under the influence of current prices, but products are sold at future prices. It means that the risk in agriculture is multiplied, due to the impossibility of correction of ongoing production. Effects of all risk factors, as well as their correlations, are focused on price. Price volatility, together with the negative effects it causes, is able to destabilize farmers' incomes, and put them off investments or optimal use of resources, which in consequence may lead to a retreat from the sector. It is crucial to provide to the agriculture sector solutions supporting farmers in taking their own economic risk. The most important thing in market stabilization seems to be its organization in both a vertical and horizontal way with special attention paid to farmers' cooperation.

Keywords Price risk · Producer groups · Market organization
Horticulture

1 Introduction

The horticultural market is a part of the agricultural market; however there are some disparities between them. What is characteristic for horticulture is the existence of many local markets, which are weakly organized. There is also greater competition

L. Zaremba (✉)
Institute of Agricultural and Food Economics, National Research Institute, Warsaw, Poland
e-mail: Lukasz.Zaremba@ierigz.waw.pl

© Springer International Publishing AG, part of Springer Nature 2018
K. Mattas et al. (eds.), *Sustainable Agriculture and Food Security*,
Cooperative Management, https://doi.org/10.1007/978-3-319-77122-9_6

among producers; seasonality is abreast of the high diversity of fruit and vegetables. Fruit and vegetables, in comparison to most agricultural products, are more vulnerable to transport conditions, and require special treatment. From the buyer's point of view, horticulture is characterized by massive intensification and diversification, as well as seasonality of consumption (Czyżewski and Kryszak 2015).

The economic outcome is a result of many factors. Due to the environmental character of horticulture production, it is hardly possible to forecast in both the short- and mid-term the economic outcome. Ongoing climate change results in various weather anomalies such as floods, droughts and heavy storms, which influence the harvest in both quality and quantity.

Moreover, globalization phenomena, typical in the 21st century economy, lead to an escalation of disparity between purchasers of horticulture products (global trade and industry) and dispersed producers. The market mechanism, which acts as a non-judgmental phenomenon, becomes an especially worrisome risk indicator (Szulce 2001). What becomes remarkably important, according to these challenges, is the issue linked with farmers' income stabilization. Nowadays farmers are exposed to a broader spectrum of external factors like never before in protective market conditions (Hamulczuk and Stańko 2008).

The regulations of the agricultural (and horticultural) market in the European Union had been embodied in the Common Agriculture Policy, in the 39th article of the Treaty of Rome. The main purpose of these regulations was to enhance agricultural efficiency by supporting technological improvements. What was also important was the rational development of agricultural production, as well as the optimal usage of production factors, especially labor, improvement of the standard of living for farmers, market stabilization, and provision of reasonable prices for agricultural products to consumers. All the goals mentioned were going to be fulfilled in view of discrepancies resulting from the geographical determinants of particular regions. Their introduction was to be gradual with the maintenance of the adjustment periods. The negotiations, which took place in Doha in 2004 by WTO representatives led to solutions towards the trade liberalization of agricultural products. What is crucial in this process is the gradual reduction of inner support, lowering the support for exports, and further openings for external partners. The reforms in The European Union tend toward the same direction, as part of the Common Agriculture Policy (CAP). Self-sufficiency, accomplished by European countries, made CAP an important tool for providing income parity of agricultural and non-agricultural production (Czyżewski and Kryszak 2015).

This paper is an overview of the processes and changes in the situation on the horticulture market in Poland. There are both EU and Polish regulations concerning the horticultural market analyzed.

2 Current Situation in EU and Poland

The situation in the horticultural market in Poland, as in the EU, is regulated by the Common Agricultural Policy (CAP). Thanks to the CAP, producers' incomes, like subsidies, are financed from taxes paid by citizens. These regulations, at least until 2013, caused cost escalation, increasing food surplus, and problems with environmental protection (Jerzak 2009). Moreover, European policy, in the scope of producers' income stabilization, used mainly interventionist tools, such as: minimal prices and withdrawal of products from the market (Filipiak 2014). Changes in the CAP after 2014 (Regulation (EU) No 1308/2013 of 17 December 2013) has moved the balance towards ecology and sustainable development. It hasn't, however, solved the problems with risk management, especially in the situation of increasing price volatility. This phenomena is a result of tight linkages between the European market and other world markets. The use of derivatives in agriculture, in order to control price risk, is rare, and in the horticulture sector it doesn't exist. The most common solution, widely used in Polish horticultural farms, is production diversification. Price risk is being disposed of in many production directions; incomes from one crop can cover losses in others. This solution is, however, inefficient. It both slows down the process of production specialization and farm technological development, and reduces the possibility of income gain (Jerzak 2009).

As an American study reveals (Czyżewski and Kryszak 2015), most agricultural processors are forward-looking; they consider their profits in the medium and long run. Smaller producers can be left behind in thin markets due to the transaction costs associated with contracting and scale economics in production (Czyżewski and Kryszak 2015) Unfavorable influences on small producers have the development of market networks. In addition, the increased coordination between producers and processors afforded by bilateral contracts reduces costs of production and opportunity costs of inputs, and transmits more information about consumer demand than the traditional cash market. Both of these outcomes increase total returns to producers and processors (Czyżewski and Kryszak 2015).

According to the Agricultural Farm Structure Survey of Poland (2013), there is a positive trend noticed in changes in the structure area of horticulture farms. There is a concentration and production specialization, especially in bigger holdings. The results of the farm structure survey in 2010 revealed a spike in orchard area by 103.2 thousand ha (38.1%), in comparison to the census of 2002. At the same time, the area of field vegetables declined by 31.8 thousand ha (18.6%), with a similar drop in the number of holdings. The results of the 2013 survey shows that in the area of permanent crops, still, orchards have the biggest share (the plantation of trees and shrubs including their nurseries). In comparison to 2010, the acreage of orchards decreased by 1.1% and the number of orchards by 28.6%. The number of fruit farms in 2013 totaled 174.8 thousand. The average orchard area increased from 1.5 ha in 2010 to 2.07 ha in 2013. Since 2010, there have also been further significant changes in the structure of fruit farm area. The fruit crops in farms with large acreage have increased. More than 60% of the total area of fruit trees were in holdings sized 5 ha and more; the plantation area of fruit bushes on the same size farms accounted for approx. 44% of the total area of fruit bushes. In comparison to 2010 orchard area with fruit trees increased by 2.0% to 267.3 thousand ha; the cultivation area of fruit bushes has grown by 2.4% to 90.8 thousand ha. Fruit trees in orchards were cultivated by 127.6 thousand farms, by 42.2 thousand (24.8%) households less than in 2010. The average orchard area and the area of fruit trees increased significantly, especially in big fruit farms. The average cultivation area of fruit trees in orchards has increased since 2010 from 1.54 to 2.10 hectares in 2013. At the same time, there was a significant decrease in the area under fruit tree cultivation in orchards of up to 1 ha inclusive. The share of this area group dropped from 14.7% in 2010 to 9.6% in 2013.

The cultivation area of field vegetables, according to 2013 outcomes was 128.2 thousand ha. In comparison to 2010 the area of field vegetables decreased by 9.6 thousand ha (6.9%). In 2013 field vegetables were grown in 90.1 thousand farms, by 1.6 thousand households (1.8%) less than in 2010, and the average area of land growing vegetables in one holding amounted to 1.42 ha—more by 5.6% in comparison to 2010. In 2013, the highest share of farms growing field vegetables was recorded in the area groups of 3–5 ha (19.4% of all households in this group) and 1–2 ha (16.8%), while the lowest was among households with an area of 100 ha and more (0.5%) (Fig. 1).

The share of fruit and vegetables in the total plant production shows no upward trend and in the years 2005–2014 it ranged around 29%, and in 2006 it amounted to 41%. In the production of food the share of processed fruit and vegetables is approx. 10–15%. Exports in the agri-food industry were slower than the growth rate of foreign sales in other industries, so the share of fruit and vegetables and their products decreases and the average in 2012–2014 amounted to 12.4%, compared to

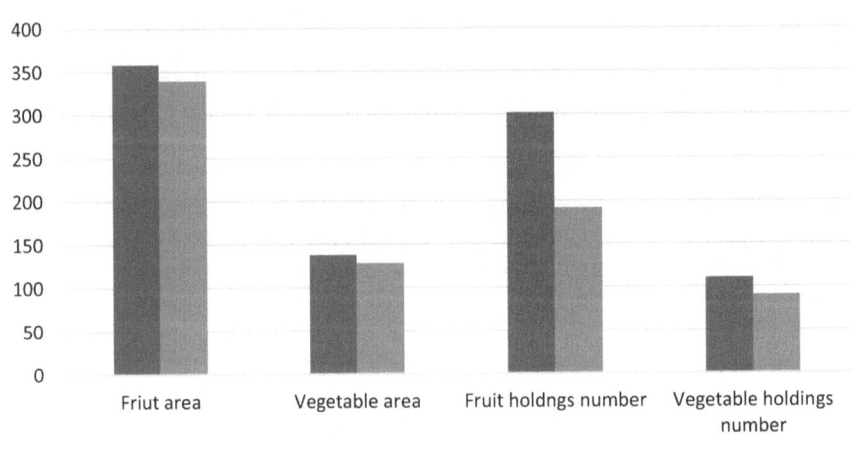

Fig. 1 Area in thousand ha, and the number of holdings in thousand ha in 2010 and 2013 in Poland *Source* Statistical Office of Poland (2010)

16.5% in 2005. There is an ongoing process in the horticultural market in Poland, towards production consolidation. Abreast with changes in the farm size structure, the number of smaller ones decreases.

3 Producer Cooperatives and Groups

Poland, while joining the EU, wasn't prepared for integration in the scope of the fruit and vegetable market organization. Despite twelve years of Polish attendance in the EU, the agriculture and horticulture market organization ratio is weak at the level of 12% while average the European organization level is about 43%. The best organized in Poland is the apple market, which is assessed at 30%.

Being one of the biggest fruit and vegetable producers in Europe, Poland makes a loss mostly because of weak cooperation between producers. There are many new solutions helping to improve producers' position in the market, but this requires enhanced cooperation and organization.

The changes observed in farm structure in Poland result from the directives of European law. Regarding this fact, producers who want to keep up with changes in the market have to strive for modernization and boost production efficiency. It is also meaningful due to the ongoing sector liberalization (globalization) that causes growing competitiveness. In many European countries, as well as in Poland, these changes are difficult to conduct, because of the still weak cooperation among producers (Jerzak 2009).

That's why the most important condition, in order to support horticultural producers, is the need for cooperation. The main goal of the group is to provide planned and adjusted demand production, especially in terms of quality and quantity. The supply concentration is to be promoted, providing products on the market. On this basis production costs are to be reduced, and as a consequence, prices will be more stable [Council Regulation (EC) No 2200/1996].

According to the latest data by the Agricultural Market Agency (AMA) (2016) published on 31 July 2016 there are 301 active groups (producers of fruits, vegetables, fruit and vegetables, herbs and mushrooms), associating 7150 members. In comparison to 2013 the total number of groups and organizations decreased. This was caused by a reduction of financial support for producer groups, which are treated as a transitional form on the way to reaching the status of a producer organization. In comparison to 2013, in 2016 there was clearly a noticeable increase in the number of recognized fruit and vegetables organizations which are the preferred form of producers' associations (Table 1).

Financial assistance is granted by the Agency for Restructuring and Modernization of Agriculture (ARMA) initially to recognized producer groups of fruit and vegetables and recognized producers of fruit and vegetable. Aid granted to producer groups allows for covering the costs of the group establishment and the administrative activity, as well as part of the eligible investment costs included in the approved plan of gaining recognition. Producers' organizations are supported

Table 1 Market organization of fruit and vegetables

	The number of groups and producer organizations		
	Total	Incl. preliminary recognition groups	Incl. producer organizations
Before integration with EU			
Until 1.05.2004	10	6	4
After integration with EU			
since 1.05.2004 to 31.12.2008	119	114	5
As at 15.07.2013	324	219	105
As at 31.07.2016	287	59	228

Source Data of Agricultural Market Agency (2016)

by additional financial support, which facilitates investments (covering 30–50% of costs). Organization members can also receive payments for products withdrawn from the market. Additional payments are not allowed to exceed 10% of the total sold production (8.5% for apple and pear). All products withdrawn from the market need to keep at least 2 class standards. Producer organizations in Poland are supported by 4.1% of members' commodity production.

Current funding for producer groups on the horizontal basis (same rules for fundraising by all groups operating in agriculture) will cause the rate of the formation of groups of fruit and vegetable to be weaker. The number of recognized organizations of fruit and vegetables will be growing at the same time. Support for these units hasn't been abolished (dedicated support for organizations operating in the fruit and vegetable sector). The organization will focus on the development of technical infrastructure and logistics organization in order to increase the concentration of supply and sales opportunities, and reduce operating costs. All the actions taken within groups are a starting point to stabilize producers' income. On the one hand, production costs are reduced and transaction costs are lower; on the other hand, crops provided by the groups is greater, in comparison to the potential of individual producers.

Most plant processors cooperate with 40–60 intermediaries. Crucial for cooperation with intermediaries is reliability in fulfilling their agreements concluded with processing plants. Most plant processors collaborate with intermediaries on the basis of reliable contracts paid on time. On the other hand, long-term cooperation with a specific group of regular suppliers is possibly powered by reliability and promptness of payments for fresh crops. This condition is both essential and sufficient to maintain a relatively stable group of intermediaries that guarantee regular and sufficient supply of fresh crop material. What is important, for producers, is not the highest price, which can be paid for crops, but payoff promptness.

4 Linkage System Between Producers and Buyers

The solution used in Polish horticulture in order to reduce price risk is vertical cooperation (Jerzak 2009). It is one of the most advanced forms of coordination, and means close cooperation on the particular steps in the production chain (Stępień and Śmigała 2012). In this situation the producer losses, which result from a weak bargaining position, are minimized. The excess which thereby occurs may be repossessed by the producer, instead of being used by the processing companies. This kind of cooperation is profitable from the producers' point of view. The companies lose, however, their privileges strong position on the market.

Supply fragmentation of fruit and vegetables hinders acquiring fresh crops for processing with the usage of contract agreements. Breaking through the barrier of the supply scale is possible thanks to private intermediary companies that are involved in the purchase; they, however, charge for profit margin amounting to 10–15% of the transaction value. The share of intermediaries in the purchase is estimated for 80%.

The small scale of vertical linkages between processing plants and fruit producers results, for the most part, from a huge purchasing price volatility of fruit; it makes it difficult to settle the level of purchase prices in contract agreements. Better price stability and lower fluctuations on the markets cause the participation of intermediaries to be low (approx. 20%) in the purchase of vegetables for processing. Contract agreements are concluded by processing plants with larger-area farms. The sale of fruit and vegetables directed to consumption in the domestic market is dominated by sales of the small and medium-sized shops, fruit and vegetable shops and groceries (approx. 35%). Sales by large retail chains are approx. 25%, wholesale markets by approx. 15%, and marketplaces approx. by 5%. Sales to retailers and wholesale markets but also in the marketplaces are carried out at more than 60% by intermediaries. Intermediaries dominate also in export sales. Direct selling on the domestic and foreign markets is carried out mainly by producer organizations and insignificantly by associations of producers' organizations and large trading companies. The sale of semi-processed products on the external markets and on the domestic market (mainly concentrated fruit juices and frozen food) is handled by big processing companies financed by foreign capital. The sales of semi-processed food by small- and medium-sized companies is being delivered by 50% abroad. Intermediaries are large production companies specialized in export companies. Sales on the domestic market are most often direct.

The weakest units in the marketing chain are fruit and vegetable producers. This concerns the cooperation with both processing plants and traders. Small- and medium-size companies are weak partners in confrontation with great processing plants which are able to act as intermediaries in exports, with foreign trade companies and large retail chains (Bożena 2016).

The useful tool that helps to organize the market and stabilize incomes is contract agreement, which is a method of purchase, based on a bilateral voluntary contract between the producer and food processor, or the purchasing company.

What is important in this agreement is to make the agreement before the production starts, so the risk of not having a buyer for crops doesn't exist. In the case of not keeping the contract conditions, there are penalties fixed by the contract (Stępień and Śmigała 2012).

Until the end of 2015 fresh fruit and vegetables were delivered to processing companies, or purchasing companies on the basis of contracts, which specified only supply quantity, quality and delivery frequency. The new regulations since 2016 have entered into force; according to them, all contracts need to be written, with specified prices and ways of fix them. Market participants, in order to adjust to the regulations, have worked out the trilateral agreement system. Processor companies sign an agreement with the middleman, who issues invoices for each delivery. This system is, however, used rarely because of system leakiness. Hence, producers still have to face real market prices. However, legislative action has been taken in the preparation of the new law, which will enforce, through criminal responsibility, the need to conclude agreements right from the start with the suppliers on prices or ways for its determination. This will change the existing system of relations and economic foundations of the processing plants, but also will limit the price competitiveness of processing entities in acquiring fresh crops. The necessity of setting purchase prices will cause loss of influence on the level of their variability depending on the situation of the global market. Contracts will still, most likely, be concluded between plant processors and intermediaries. Intermediaries will require purchase price agreements with producers. Nevertheless, plant processors will be much more involved in the process of prices establishing.

5   Insurances

Insurance of horticultural production is an agreement between producers and an insurance company that allows the producer, after the regulation of an insurance premium, to claim damages (Stępień and Śmigała 2012). Voluntary insurances are unpopular in Poland, because of farmers' aversion to bearing the relatively high costs. On the other hand, the insurance offer is unattractive and includes many exclusions. National support for insurance for horticulture is limited. There are only three insurance companies that offer products for agriculture and horticulture. Besides, from 1 July 2008, the producer is obliged to conclude a compulsory insurance agreement of crops against the risk of drought, hail, flood, negative effects of storage and spring frosts on at least 50% of the crop (crop insurance act for agriculture and livestock from 7 July 2005). From 1 January 2010 in all 27 EU countries, farmers who do not include insurance for at least half their production (crops) in case of disasters will receive government support reduced by half (EC Regulation 1857 of 2006, 57 of 2006). There is however no price risk insurance on the Polish market, while in more developed countries, like the USA and Canada, such solutions are well known.

6 Conclusions

All issues concerned with price risk limitations in the horticultural sector are rather characteristic for developed countries. What is crucial for farmers' income stabilization is for the most part good market organization. This process is slow in Poland, but since the beginning of Polish existence in the European structures, it is ongoing. In comparison to other countries from the EU-13, market organization ratio in Poland is weak. The producers are those who in the marketing chain of fruit and vegetable deliveries have the weakest position. This results from the considerable farm fragmentation, and the weak supply offer. That is why, especially for non-union producers, they have to accept the offered prices by processing companies and retailers. There is no straight support in Poland for producers in the scope of price protection, but there are administrative tools and solutions which help to prevent income loss. In order to improve the position of Polish farmers on the market, more attention should be paid on economic education; it would help to act on the increasingly more liberal market. The cognizance of threats and the ability of the use of economic tools and indicators will foster taking economic responsible decisions.

References

Adjemian, M. K., Brorsen, B. W., Hahn, W., Saitine, T. L., & Sexton, R. J. (2016). Thining markets in U.S. agriculture. brak miejsca: United States Departament of Agriculture.
Agricultural Farm Structure Survey of Poland in 2010. (2010).
Agricultural Farm Structure Survey of Poland in 2013. (2013).
Bożena, N. (2016). *Rynek owoców i warzyw w Polsce*. Warszawa: IERIGŻ.
Czyżewski, A., & Kryszak, Ł. (2015). Relacje cenowe w rolnictwie polskim a dochodowość gospodarstw rolnych i gospodarstwa domowych rolników. Warszawa: Zeszyty Naukowe SGGW w Warszawie, *Problemy Rolnictwa Światowego strony 15*(3), 17–29.
Data of Agricultural Market Agency (AMA). (2016).
Data of Statistical Office of Poland (GUS). (2010).
Filipiak, T. (2014). *Zmiany na rynku warzyw i w gospodarstwach warzywniczych w Polsce po integracji z Unią Europejską*. Warszawa: SGGW.
Hamulczuk, M., & Stańko, S. (2008). *Zarządzanie ryzykiem cenowym a możliwości stabilizowania dochodów producentów rolnych*. Warszawa: Instytut Ekonomiki Rolnictwa i Gospodarki Żywnościowej-Państwowy Instytut Badawczy.
Jerzak, M. (2009). *Zarządzanie ryzykiem cenowym jako zynnik poprawy konkurencyjności gospodarstw rolnych w warunkach liberalizacji Wspólnej Polityki Rolnej w UE*. Poznań: Uniwersytet Poznański.
Sobczak, W., & Wielechowski, M. (2016). Rozwój grup i organizacji producentów owoców i warzyw w Polsce w świetle unijnego i krajowego ustawodawstwa, Warszawa.
Stępień, S., & Śmigała, M. (2012). Price risk management in agriculture in practice of selected countries in the world. Warszawa: SGGW.
Szulce, H. (2001). *Uwarunkowania i możliwości sterowania ryzykiem w produkcji rolnej*. Poznań: Wydawnictwo Akademii Ekonomicznej w Poznaniu.

Willingness to Pay for Malaria Prophylaxis in Ethiopia

Simon O. Soname and Garth J. Holloway

Abstract In this chapter, we measure how much malaria impacts on farmers' technical efficiency and the application of these values to present a reliable measure of the farmers' Willingness-To-Pay for malaria abatement in Ethiopia. Malaria is one of the diseases that has prevented the African continent from achieving its main goal of food availability, security and sustainable development. One major problem policy makers seem to face all the time is the precise amount the households are willing to pay for a prophylactic measure. The epidemiology of the disease on the continent has made this measure difficult and adopting a stated preference approach has not been very helpful. Also, the link between malaria incidence and agricultural productivity has not been fully explored in the literature. We use a dataset from Ethiopia with the corresponding spatial malaria prevalence dataset from the Malaria Atlas Project. We apply this dataset to the envelope theorem to arrive at a reliable estimate of the Willingness-To-Pay and a measure of how much malaria affects farmers' technical efficiency. The merger of the household dataset, with the spatial malaria dataset and the innovative use of the envelope theorem, is one of the major high points of this chapter. We apply Bayesian Econometrics to our empirical framework. The results show that in Ethiopia, malaria affects efficiency and has an a priori sign. The results further state that for a 100-unit increase in malaria, the household is willing to pay, on average, US$0.12 to purchase prophylactic measures. Policy makers can use these values to introduce minimum prices and gradual repayment schemes for prophylactic measures.

Keywords Willingness-to-pay · Household model · Roy's identity
Bayesian analysis

S. O. Soname (✉) · G. J. Holloway
Department of Agri-Food Economics, University of Reading, Reading RG6 6AR, UK
e-mail: ssoname@hotmail.com

G. J. Holloway
e-mail: garth.holloway@reading.ac.uk

© Springer International Publishing AG, part of Springer Nature 2018
K. Mattas et al. (eds.), *Sustainable Agriculture and Food Security*,
Cooperative Management, https://doi.org/10.1007/978-3-319-77122-9_7

1 Introduction

In the year 2000, Africa's heads of government convened in Abuja, Nigeria's capital to chart a new course for the control and possible eradication of malaria from the continent (Nigeria Ministry of Health 2001). At the end of the conference they came out with a communique otherwise called "The Abuja Declaration", of which the main target was the halving of malaria by the year 2010. This declaration is in line with the recently launched Sustainable Development Goals in 2015.

Insecticide treated nets distribution, indoor residual spraying, distribution of free anti-malaria drugs, and maintaining a clean environment are part of the measures that the Roll Back Malaria Partnership intend to use to achieve the aims and objectives of the "Abuja Declaration". The use of nets to prevent nuisance bites from mosquitoes has been in use for a long time, but they have not been effective in preventing/controlling the disease as they only serve as physical barriers against the vector. The insecticide treated nets are bed nets impregnated with pyrethroid chemicals and serve not only as a physical barrier between man and the vector but also have an "excite-repellent" effect which serves as a chemical barrier between man and his host (WHO 2007).

Malaria is a vector-borne disease caused by the parasite Plasmodium and its vector is the mosquito. There are different species of Plasmodium but the common ones in Africa are *Plasmodium falciparum* and *Plasmodium vivax*. More than two billion people live at risk of contracting malaria in the world with over 300 million clinical cases of the disease per year. These two species are also common in Ethiopia. According to the World Health Organisation (WHO 2016), about 75% of the total land area in Ethiopia is prone to malaria which exposes about 60% of the population to scourge of the disease. It constitutes about 14% of outpatient visits in Ethiopia (USAID and The Ethiopian Government 2011).

Increasing agricultural productivity is sine qua non for Africa's survival. In fact, increasing agricultural productivity forms a great part of the policy strategy and manifestos of most African governments, but it seems that to attain this goal is elusive. Measuring the impact of malaria is a somewhat delicate process because unlike other diseases like HIV/AIDS that have similar effects across age ranges, malaria's effect across different age ranges is quite varied. Also, the disease tends to manifest itself at the peak of the planting season where labour demand is quite high and at a time where hired labour might not be a perfect substitute for family labour for farm activities.

Good health and productive agriculture are minimum requirements in the economy of a nation in the fight against poverty and in the development of its human capital as health capital is an indispensable input in agricultural production. Information on the impact of malaria on farming households' efficiency is important so as to target intervention efficiently and equitably and to justify investment in research and control.

From the foregoing, it is evident that the ability to measure and improve the productive efficiency of the farming household is critical to alleviating poverty in Africa.

The parametric measurement of productivity and efficiency is based on the use of different functions.[1] These are: the production function (for example: Welsch 1965; Chennareddy 1967; Afriat 1972; Battese et al. 1989; Battese and Coelli 1995; Cullinane et al. 2006), the cost function (Kopp and Diewert 1982; Zieschang 1983; Asche et al. 2009; Dong et al. 2014); the revenue function (Diewert 1974; Berger et al. 1996); the profit function (Lau 1972; Lopez 1985; Kumbhakar 2001; Humphrey and Pulley 1997; Alvarez and Arias 2004; Ray and Das 2010); and finally, the distance function[2]—used for multi-input-output efficiency analysis (Shepard 1953; Feng and Serletis 2010; Newman and Matthews 2006; Siegel et al. 2008; Riccardi et al. 2012; Fa" re and Grosskopf 2010).

These functions form the foundation of the stochastic frontier model,[3,4] (also called the composed error model) used in the parametric analysis of productivity and efficiency. The model was developed simultaneously by Aigner et al. (1977), Meeusen and van den Broeck (1977), and Battese and Corra (1977). It has seen application in agriculture (Coelli et al. 2003; Wadud and White 2000; Amaza and Olayemi 2002; Ajibefun and Abdulkadri 1999; Balcombe et al. 2006; Ogundele and Okoruwa 2006; Amos 2007; Jara-rojas et al. 2012); in public policy and evaluation (Seyoum et al. 1998; Abdulai and Huffman 2000; Ajibefun 2002; Abdulai and Eberlin 2001; Areal et al. 2012); in livelihood and fisheries (Bezemer et al. 2005; Holloway et al. 2005; Holloway and Tomberlin 2007; Tomberlin and Holloway 2008); in health (Croppenstedt and Muller 2000; Ajani and Ugwu 2008; Ajani and Ashagidigbi 2008; Loureiro 2009; Ulimwengu 2008); and, in spatial studies and migration (Tadesse and Krishnamoorthy 1997; Wang and Schmidt 2002; Chen et al. 2009).

Apart from the use of the stochastic frontier model in developing our conceptual framework, we also utilize the household model (because this paper focuses on the household); Singh et al. (1986) extensively discuss its use. The household model also sees vast application in agriculture and health (Yotopoulos et al. 1976; Simtowe and Zeller 2006; Ulimwengu 2009; Badiane and Ulimwengu 2013; Muyanga and Jayne 2014a); in transaction cost (Omamo 1998; Lofgren and Robinson 1999; Key et al. 2000); in energy (Heltberg et al. 2000; Chen et al. 2006; Muller and Yan 2016); in demand and consumption (Browning and Meghir 1991; Saha and Stroud 1994; Chavas 2006; Muyanga and Musyoka 2014b; Dalton 2004; Doiron and Kalb 2004); in labour supply and time allocation (Huffman 1980;

---

[1]The literature uses 'frontier' and 'function' interchangeably with both meaning the same, whereas 'model' is occasionally used in the literature to mean the same as 'function'. In this case it is different, as 'model' derives from functions or a modification of a function.

[2]The distance function is the root of all of the aforementioned functions. Thus, each of the aforementioned functions derives from the distance function.

[3]See Murillo-Zamorano (2004), Soname (2016) for a thorough review of the model.

[4]Farrell (1957) dichotomises *overall* efficiency into technical and allocative efficiency respectively.

Huffman and Lange 1989; Skoufias 1994; Goodwin and Holt 2002; El-Osta et al. 2008; Nepal and Nelson 2015); and finally, in bio-economics (De Janvry and Sadoulet 1995; Ruben and van Ruijven 2001; Kebede et al. 2016).

Thus, the main objective of this paper is to estimate the Willingness–To–Pay for malaria abatement in Ethiopia through the assessment of the technical efficiency and productivity of the farmers. The remaining part of this paper is arranged as follows: we discuss the data and analytical methods in Sect. 2, and the results and conclusion in Sects. 3 and 4 respectively.

2   Data and Methods

We utilize data from the 1994 version of the panel survey collected by the International Food Policy Research Institute (IFPRI) and the University of Oxford. The 1994 dataset focuses on the collection of data from diverse farming households in different regions in Ethiopia cultivating different crops under different field conditions. The total number of peasant associations surveyed (villages) is fifteen.

The other data we employed was the spatial malaria data from the Malaria Atlas Project. We merged these two data together in order to obtain the data for our analysis. This data was applied to our conceptual framework which we state as:

$$\Delta m \equiv p_a \cdot f(L, A, X) \cdot \theta_j^{\prime *} \cdot \Delta \theta \tag{1}$$

where Δm is the willingness of the farmer to pay for malaria abatement, p_a is the price of agricultural staples, and $f(L, A) \cdot \theta_j^{\prime *}$ is the composed error model with L, A, X being labour, size of land used and other factors that the researcher conjectures will affect the frontier part of the model. The quantity, $\theta_j^{\prime *}$, is a factor that affects the inefficiency: we input the malaria variable in this part of the composed error model. The choice of where to input the malaria variable in the composed error model is based on the results of a model selection algorithm which is carried out prior to estimation. The quantity, $\Delta \theta$, is a change in malaria prevalence in Ethiopia whose value is fixed by the researcher. The willingness to pay is thus a product of price of staples, the malaria prevalence estimate and the $\Delta \theta$ value.

The Bayesian method is applied to the composed error model using proper prior. The Markov Chain Monte Carlo methods of Gibbs sampling and the Metropolis Hastings were employed during our analysis. We carried out the Markov chain Monte Carlo diagnostics to decide how long we have to run our code for. The Raftery-Lewis diagnostics was employed. As stated earlier, to decide on which side of the model the malaria variable should be inputted, we carry out a model selection algorithm.

3 Results

The result of the Raftery-Lewis diagnostics shows that we do not need a burn-in. It shows that the code starts converging from one thousand draws. We ran the algorithm on thirteen variables: land area cultivated, labour, fertiliser, land type, index of other input, asset, credit, slope, intercropping, age, education, malaria, and finally, the constant. We commence by presenting the covariate selection results from the first phase of the exercise in Figs. 1 and 2, followed by the model selection results from Figs. 3, 4 and 5 and then the composed error results in Fig. 6 and Table 1.

Using a criterion of ≥ 0.05, the selected variables are: *land area, fertilizer, land type, asset, credit, slope, intercropping, malaria, constant.*

Next, we present the results of the model selection exercise. We display the results of the composed-error model with the malaria variable on the inefficiency part of the model. This is because it has the highest marginal likelihood estimate (of 1406) of the four models in the second phase of our analysis.

On looking at Figs. 3 and 4, it can be seen that the malaria variable on the inefficiency part of the model has the highest probability; thus, we place malaria on the inefficiency part of the model.

The frontier part of Table 1 shows that apart from asset, all the variables are significant at 5%. Thus, they significantly affect productivity. Fertiliser and asset are positive, thus an increase in these variables will cause an increase in productivity, other things being equal and vice versa. Intercropping and slope are negatively related to productivity. Thus, any increase in any of these variables will cause

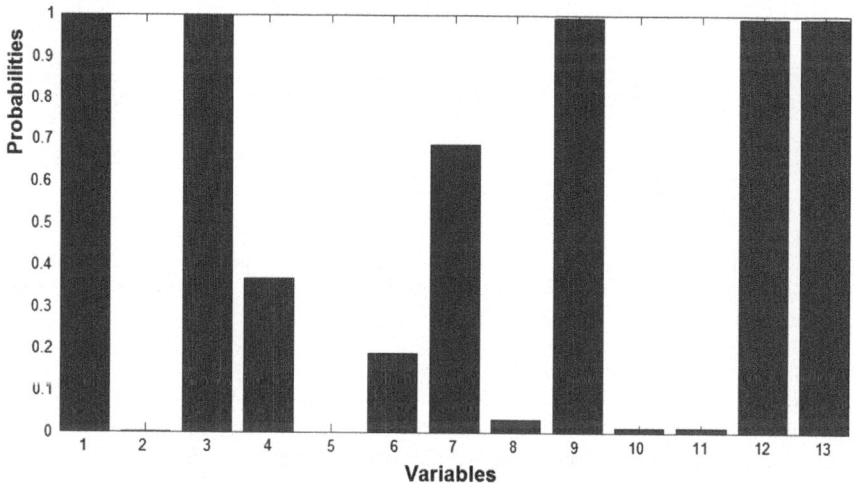

Fig. 1 Covariate selection for the Ethiopia data. 1. Landarea 2. Labour 3. Fertiliser 4. Landtype 5. Otherinput 6. Asset 7. Credit 8. Slope 9. Intercropping 10. Age 11. Education 12. Malaria 13. Constant

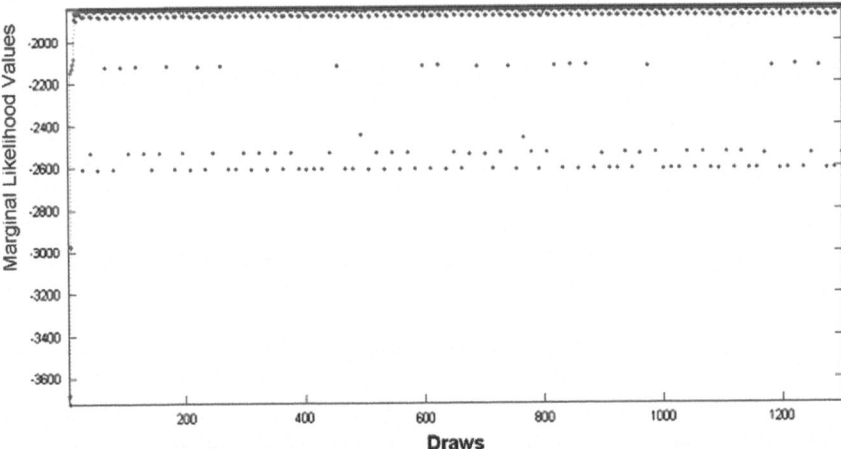

Fig. 2 Marginal likelihood estimates of the Ethiopia data covariate selection model. Red lines are the accepted densities and blue lines are the proposal densities

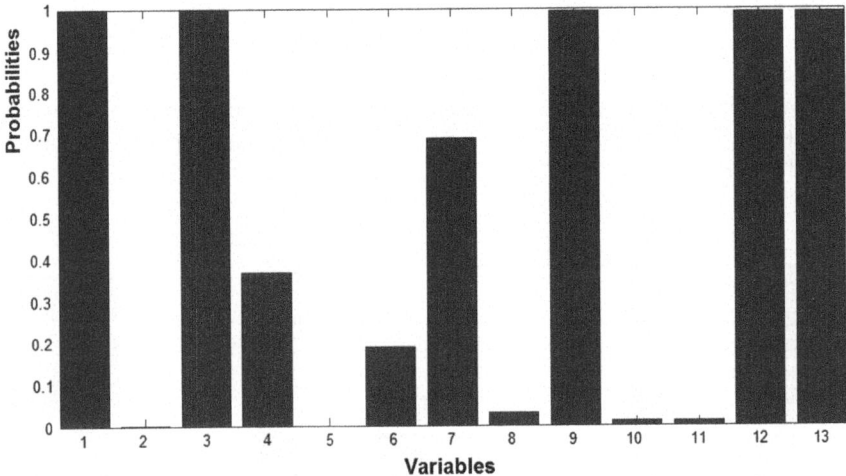

Fig. 3 Ethiopia model selection result for the *frontier* part of the composed-error model. 1. Landarea 2. Labour 3. Fertiliser 4. Landtype 5. Otherinput 6. Asset 7. Credit 8. Slope 9. Intercropping 10. Age 11. Education 12. Malaria 13. Constant

productivity to decline. The result for intercropping is quite surprising because it is expected to increase productivity. This may be because the farmers plant on the same land similar crops that require similar soil nutrients.

All of our variables in the inefficiency part of the model are significant including the malaria variable at 5%. Apart from the constant term, all the variables are positive. The sign for our malaria variable follows the expected outcome (although

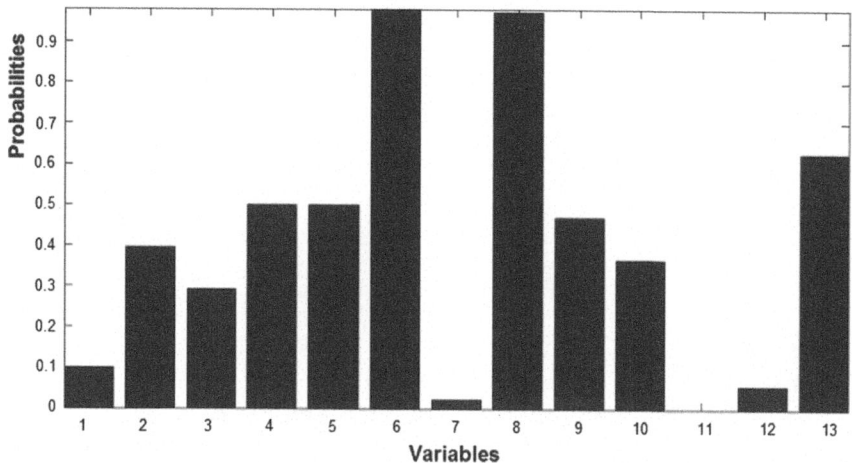

Fig. 4 Ethiopia model selection result for the *inefficiency* part of the composed-error model. 1. Landarea 2. Labour 3. Fertiliser 4. Landtype 5. Otherinput 6. Asset 7. Credit 8. Slope 9. Intercropping 10. Age 11. Education 12. Malaria 13. Constant

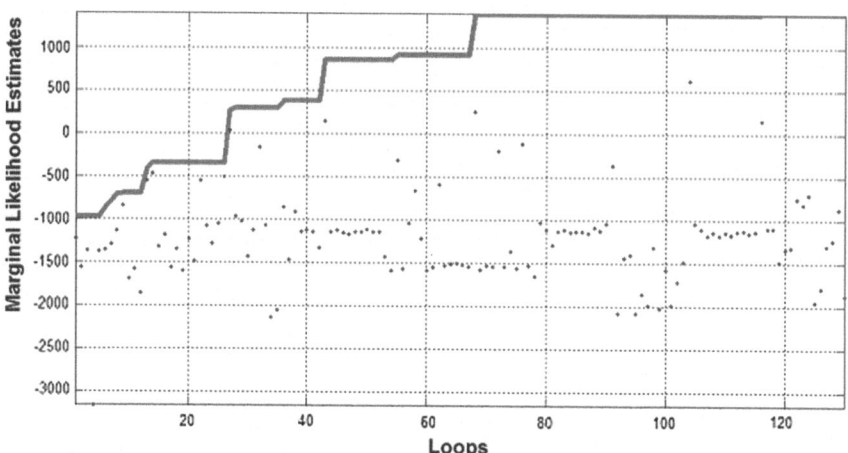

Fig. 5 Marginal likelihood estimate of the model with *Malaria* on the inefficiency part of the composed-error model. *We produced the mean of ten loops*

the signs for the other variables did not follow the expected outcome as we expect them to be negatively related to inefficiency). Our results shows that an 81% increase in malaria cases per 1000 per annum will cause a 100% increase in inefficiency and vice versa.

On multiplying the price of staples by the malaria prevalence estimate, we obtain a willingness to pay value of about 11 Ethiopian birr (about US$0.98 at the 2009 conversion rate) if malaria increases by 100% cases per 1000 individuals per

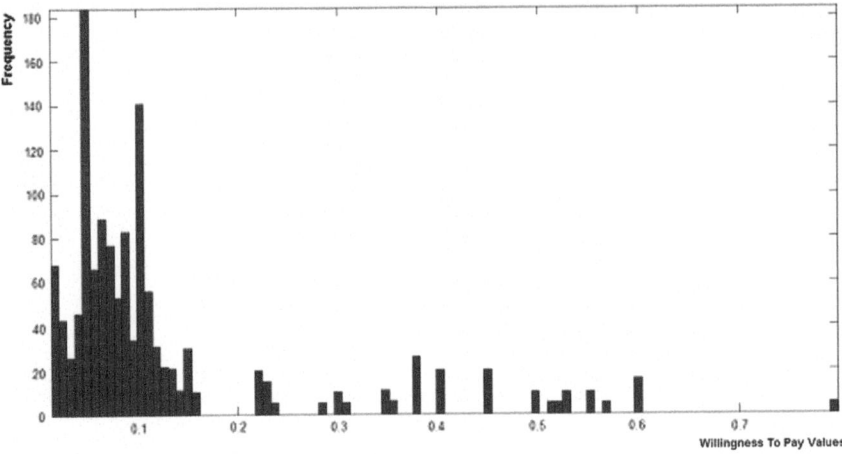

Fig. 6 Ethiopia Willingness-To-Pay posterior distribution (values in Ethiopia Birr)

annum. The posterior distribution in Fig. 6 is positively skewed and shows that most farmers in the distribution are willing to pay between 0 Birr (US$0) and about 15 Ethiopian Birr (about US$1.34) if malaria increases by 100% cases per 1000 individuals per annum. The analysis shows a high regression coefficient value of

Table 1 Ethiopia estimates of the composed error model

S/N	Variables	Ethiopia	
		Mean	95% highest posterior density
Frontier variables			
1	Fertiliser	1.44[a]	[0.97, 1.92]
2	Asset	0.50	[−0.29, 1.33]
3	Slope	−0.44[a]	[−0.79, −1.08]
4	Intercropping	−0.73[a]	[−1.061 −0.406]
5	Constant	7.13[a]	[6.65, 7.62]
Inefficiency variables			
6	Land area	3.31[a]	[2.56 3.86]
7	Fertiliser	2.41[a]	[1.46 3.19]
8	Credit	0.45[a]	[0.22 0.64]
9	Malaria	0.81[a]	[0.47 1.11]
10	Constant	−0.78[a]	[−1.16 −0.53]
21	wtpay	0.11	N/A
25	R2	0.79	N/A
26	Average eff.score	0.89	N/A
27	Mean predic. val.	6.14	N/A

[a]Significant at 95% highest posterior density interval/level of significance

79%. This means our model fits the data very well. The mean predicted value is about 6.14 kg ha^{-1}.

Our results reveal that Ethiopian farmers have a mean efficiency score of about 90%. This assertion supports Schultz's hypothesis, that peasant farmers are poor but efficient.

4 Conclusion

We have attempted to investigate the significance of malaria in productivity and efficiency measurement in Ethiopia. We conjectured that malaria should influence the farmers' efficiency more; the results from the analysis of the Ethiopian data supports our position. However, care needs to be taken in generalising this finding as the literature is still not clear about this. Our use of the Bayesian covariate selection algorithm and the Markov chain diagnostics is a promising way of reaching a universal consensus.

On average, farmers in our entire sample are willing to pay less than one United States dollar for a 100% increase in malaria cases per 1000 individuals per annum. This result has implications for policy making as it will help determine the nature and level of subsidy required for whatever prophylactic measure chosen to prevent/reduce malaria.

Acknowledgements Copies of the computer codes are available from Simon Soname at: ssoname@hotmail.com. The authors would like to thank the Malaria Atlas Project and the World Bank for providing the data used in this reserach. We also thank the anonymous referees for their useful comments. All errors are our responsibility.

References

Abdulai, A., & Eberlin, R. (2001). Technical efficiency during economic reform in Nicaragua: Evidence from farm household survey data. *Economic Systems, 25*(2), 113–125. http://www.sciencedirect.com/science/article/pii/S0939362501000103.

Abdulai, A., & Huffman, W. (2000). Structural adjustment and economic efficiency of rice farmers in northern Ghana. *Economic Development and Cultural Change, 48*(3), 503–520. http://www.jstor.org/stable/10.1086/452608.

Afriat, S. N. (1972). Efficiency estimation of production functions. *International Economic Review, 13*(3), 568–598. http://www.jstor.org/stable/2525845.

Aigner, D., Lovell, C., & Schmidt, P. (1977). Formulation and estimation of stochastic frontier production function models. *Journal of Econometrics, 6,* 21–37.

Ajani, O., & Ashagidigbi, W. (2008). Effect of malaria on rural households' farm income in Oyo state, Nigeria. *African Journal of Biomedical Research 11.*

Ajani, O., & Ugwu, P. (2008). Impact of adverse health on agricultural productivity of farmers in kainji basin north-central Nigeria using a stochastic production fron tier approach. *Trends In Agricultural Economics, 1*(1), 1–7.

Ajibefun, I. A. (2002). Analysis of policy issues in technical efficiency of small scale farmers using the stochastic frontier production function: With application to Nigerian farmers. http:// ageconsearch.umn.edu/bitstream/7015/2/cp02aj01.pdf.

Ajibefun, I. A., & Abdulkadri, A. O. (1999). An investigation of technical inefficiency of production of farmers under the national directorate of employment in Ondo state, Nigeria. *Applied Economics Letters, 6*(2), 111–114.

Alvarez, A., & Arias, C. (2004). Technical efficiency and farm size: A conditional analysis. *Agricultural Economics, 30*(3), 241–250. http://www.sciencedirect.com/science/article/pii/ S0169515004000246.

Amaza, P. S., & Olayemi, J. K. (2002). Analysis of technical inefficiency in food crop production in Gombe state, Nigeria. *Applied Economics Letters, 9*(1), 51–54.

Amos, T. (2007). An analysis of productivity and technical efficiency of smallholder cocoa farmers in Nigeria. *Journal of Social Sciences, 15*(2), 127–133.

Areal, F. J., Tiffin, R., & Balcombe, K. (2012). Farm technical efficiency under a tradable milk quota system. *Journal of Dairy Science, 95*(1), 50–62.

Asche, F., Roll, K. H., & Tveteras, R. (2009). Economic inefficiency and environmental impact: An application to aquaculture production. *Journal of Environmental Economics and Management, 58*(1), 93–105.

Badiane, O., & Ulimwengu, J. (2013). Malaria incidence and agricultural efficiency in Uganda. *Agricultural Economics, 44*(1), 15–23.

Balcombe, K., Fraser, I., & Kim, J. H. (2006). Estimating technical efficiency of Australian dairy farms using alternative frontier methodologies. *Applied Economics, 38*(19), 2221–2236.

Battese, G. E., & Coelli, T. J. (1995). A model for technical inefficiency effects in a stochastic frontier production function for panel data. *Empirical Economics, 20*(2), 325–332.

Battese, G. E., Coelli, T. J., & Colby, T. (1989). *Estimation of frontier production functions and the efficiencies of Indian farms using panel data from ICRISAT's village level studies*, Department of Econometrics, University of New England.

Battese, G. E., & Corra, G. S. (1977). Estimation of a production frontier model: With application to the pastoral zone of eastern Australia. *Australian Journal of Agricultural and Resource Economics, 21*(3), 169–179.

Berger, A. N., Humphrey, D. B., & Pulley, L. B. (1996). Do consumers pay for one-stop banking? Evidence from an alternative revenue function. *Journal of Banking & Finance, 20*(9), 1601–1621.

Bezemer, D., Balcombe, K., Davis, J., & Fraser, I. (2005). Livelihoods and farm efficiency in rural georgia. *Applied Economics, 37*(15), 1737–1745.

Browning, M., & Meghir, C. (1991). The effects of male and female labor supply on commodity demands. *Econometrica: Journal of the Econometric Society* 925–951.

Chavas, J.-P. (2006). Dynamic considerations in nutrition and food demand.

Chen, L., Heerink, N., & van den Berg, M. (2006). Energy consumption in rural China: A household model for three villages in Jiangxi province. *Ecological Economics, 58*(2), 407–420.

Chen, Z., Huffman, W. E., & Rozelle, S. (2009). Farm technology and technical efficiency: Evidence from four regions in China. *China Economic Review, 20*(2), 153–161.

Chennareddy, V. (1967). Production efficiency in South Indian agriculture. *Journal of Farm Economics, 49*(4), 816–820. http://www.jstor.org/stable/1236938.

Coelli, T., Rahman, S., & Thirtle, C. (2003). A stochastic frontier approach to total factor productivity measurement in Bangladesh crop agriculture, 1961–92. *Journal of International Development, 15*(3), 321–333.

Croppenstedt, A., & Muller, C. (2000). The impact of farmers' health and nutritional status on their productivity and efficiency: Evidence from Ethiopia. *Economic Development and Cultural Change, 48*(3), 475–502. http://www.jstor.org/stable/10.1086/452607.

Cullinane, K., Wang, T.-F., Song, D.-W., & Ji, P. (2006). The technical efficiency of container ports: Comparing data envelopment analysis and stochastic frontier analysis. *Transportation Research Part A: Policy and Practice, 40*(4), 354–374.

Dalton, T. J. (2004). A household hedonic model of rice traits: economic values from farmers in West Africa. *Agricultural Economics, 31*(2–3), 149–159.

De Janvry, A., & Sadoulet, E. (1995). Household modeling for the design of poverty alleviation strategies. *Revue Economie du Developpement, 3*, 3–23.

Diewert, W. E. (1974). Functional forms for revenue and factor requirements functions. *International Economic Review*, 119–130.

Doiron, D., & Kalb, G. (2004). Demands for childcare and household labour supply in Australia. *Economic Records, 81*(254), 215–236.

Dong, Y., Hamilton, R., & Tippett, M. (2014). Cost efficiency of the Chinese banking sector: A comparison of stochastic frontier analysis and data envelopment analysis. *Economic Modelling, 36*, 298–308.

El-Osta, H. S., Mishra, A. K., & Morehart, M. J. (2008). Off-farm labor participation decisions of married farm couples and the role of government payments. *Applied Economic Perspectives and Policy, 30*(2), 311–332. http://aepp.oxfordjournals.org/content/30/2/311.abstract.

Färe, R., & Grosskopf, S. (2010). Directional distance functions and slacks-based measures of efficiency. *European Journal of Operational Research, 200*(1), 320–322.

Farrell, M. J. (1957). The measurement of productive efficiency. *Journal of the Royal Statistical Society. Series A (General), 120*(3), 253–290. http://www.jstor.org/stable/2343100.

Federal Ministry of Health, Nigeria. (2001). *The Abuja declaration and the plan of action*. Abuja. http://www.rollbackmalaria.org/microsites/wmd2011/abuja_declaration_final.html.

Feng, G., & Serletis, A. (2010). Efficiency, technical change, and returns to scale in large us banks: Panel data evidence from an output distance function satisfying theoretical regularity. *Journal of Banking & Finance, 34*(1), 127–138.

Goodwin, B. K., & Holt, M. T. (2002). Parametric and semiparametric modeling of the off-farm labor supply of agrarian households in transition Bulgaria. *American Journal of Agricultural Economics*, 184–209.

Heltberg, R., Arndt, T. C., & Sekhar, N. U. (2000). Fuelwood consumption and forest degradation: A household model for domestic energy substitution in rural India. *Land Economics, 76*(2), 213–232. http://www.jstor.org/stable/3147225.

Holloway, G., & Tomberlin, D. (2007). Bayesian ranking and selection of fishing boat efficiencies. *Marine Resource Economics, 21*, 415–432.

Holloway, G., Tomberlin, D., & Irz, X. (2005). Hierarchical analysis of production efficiency in a coastal trawl fishery. In R. Scarpa & A. Alberini (Eds.), *Applications of simulation methods in environmental and resource economics in The economics of non-market goods and resources* (Vol. 6, pp. 159–185). Netherlands: Springer.

Huffman, W. E. (1980). Farm and off-farm work decisions: The role of human capital. *The Review of Economics and Statistics, 62*(1), 14–23. http://www.jstor.org/stable/1924268.

Huffman, W., & Lange, M. D. (1989). Off-farm work decisions of husbands and wives: Joint decision making. *The Review of Economics and Statistics, 71*(3), 471–480.

Humphrey, D. B., & Pulley, L. B. (1997). Banks' responses to deregulation: Profits, technology, and efficiency. *Journal of Money, Credit, and Banking*, 73–93.

Jara-rojas, R., Bravo-Ureta, B. E., Moreira, V., & Da~az, J. (2012). Natural resource conservation and technical efficiency from small-scale farmers in central Chile.

Kebede, E., Gan, J., & Kagochi, J. M. (2016). Agriculture based energy for rural house-hold income and well-being: East African experience. *Renewable and Sustainable Energy Reviews, 53*, 1650–1655.

Key, N., Sadoulet, F., & Janvry, A. D. (2000). Transactions costs and agricultural house-hold supply response, *American Journal of Agricultural Economics, 82*(2), 245–259. http://ajae.oxfordjournals.org/content/82/2/245.abstract.

Kopp, R. J., & Diewert, W. E. (1982). The decomposition of frontier cost function deviations into measures of technical and allocative efficiency. *Journal of Econometrics, 19*(2–3), 319–331.

Kumbhakar, S. C. (2001). Estimation of profit functions when profit is not maximum. *American Journal of Agricultural Economics, 83*(1), 1–19.

Lau, L. J. (1972). Profit functions of technologies with multiple inputs and outputs. *The Review of Economics and Statistics, 281*–289.

Lofgren, H., & Robinson, S. (1999). Nonseparable farm household decisions in a computable general equilibrium model. *American Journal of Agricultural Economics, 81*(3), 663–670.

Lopez, R. E. (1985). Structural implications of a class of flexible functional forms for profit functions. *International Economic Review*, 593–601.

Loureiro, M. L. (2009). Farmers' health and agricultural productivity. *Agricultural Economics, 40* (4), 381–388.

Meeusen, W., & van den Broeck, J. (1977). Efficiency estimation from Cobb-Douglas production functions with composed error. *International Economic Review, 18*(2), 435–444. http://www.jstor.org/stable/2525757.

Muller, C., & Yan, H. (2016). Household fuel use in developing countries: Review of theory and evidence.

Murillo-Zamorano, L. R. (2004). Economic efficiency and frontier techniques. *Journal of Economic Surveys, 18*(1), 33–77.

Muyanga, M., & Jayne, T. (2014a). Effects of rising rural population density on small-holder agriculture in Kenya. *Food Policy, 48*, 98–113.

Muyanga, M., & Musyoka, P. (2014b). Household incomes and poverty dynamics in rural kenya: A panel data analysis. *Journal of Poverty Alleviation and International Development, 5*(2), 43–76.

Nepal, A., Nelson, C. et al. (2015). Estimation of shadow wage in agricultural household model in Nepal. In: *2015 Conference, August 9–14, 2015, Milan, Italy*, number 212527. International Association of Agricultural Economists.

Newman, C., & Matthews, A. (2006). The productivity performance of Irish dairy farms 1984–2000: A multiple output distance function approach. *Journal of Productivity Analysis, 26*(2), 191–205.

Ogundele, O., & Okoruwa, V. (2006). Technical efficiency differentials in rice production technologies in Nigeria. *Research Paper 154*, AERC, Nairobi. http://www.aercafrica.org/documents/rp154.pdf.

Omamo, S. W. (1998). Farm to market transaction costs and specialisation in small scale agriculture: Explorations with a non-separable household model. *The Journal of Development Studies, 35*(2), 152–163.

Ray, S. C., & Das, A. (2010). Distribution of cost and profit efficiency: Evidence from Indian banking. *European Journal of Operational Research, 201*(1), 297–307.

Riccardi, R., Oggioni, G., & Toninelli, R. (2012). Efficiency analysis of world cement industry in presence of undesirable output: Application of data envelopment analysis and directional distance function. *Energy Policy, 44*, 140–152.

Ruben, R., & van Ruijven, A. (2001). Technical coefficients for bio-economic farm house-hold models: A meta-modelling approach with applications for southern Mali. *Ecological Economics, 36*(3), 427–441. http://www.sciencedirect.com/science/article/pii/S0921800900002408.

Saha, A., & Stroud, J. (1994). A household model of on-farm storage under price risk. *American Journal of Agricultural Economics, 76*(3), 522–534.

Seyoum, E. T., Battese, G. E., & Fleming, E. M. (1998). Technical efficiency and productivity of maize producers in eastern Ethiopia: A study of farmers within and outside the Sasakawa-global 2000 project. *Agricultural Economics, 19*(3), 341–348. http://www.sciencedirect.com/science/article/pii/S0169515098000371.

Shepard Ronald, W. (1953). *Cost and production functions*. Princeton: Princeton University Press.

Siegel, D., Wright, M., Chapple, W., & Lockett, A. (2008). Assessing the relative performance of university technology transfer in the US and UK: A stochastic distance function approach. *Economics of Innovation and New Technology, 17*(7–8), 717–729.

Simtowe, F., & Zeller, M. (2006). The impact of access to credit on the adoption of hybrid maize in Malawi: An empirical test of an agricultural household model under credit market failure.

Singh, I., Squire, L., & Strauss, J. (1986). *Agricultural household models—Extension, applications and policy*. Baltimore and London: For the World Bank: The John Hopkins University Press.

Skoufias, E. (1994). Using shadow wages to estimate labor supply of agricultural house-holds. *American Journal of Agricultural Economics, 76*(2), 215.

Soname, S. O. (2016). Effects of malaria on farmers' technical efficiency in Africa. Ph.D. thesis at the University of Reading, UK.

Tadesse, B., & Krishnamoorthy, S. (1997). Technical efficiency in paddy farms of Tamil Nadu: An analysis based on farm size and ecological zone. *Agricultural Economics, 16*(3), 185–192. http://www.sciencedirect.com/science/article/pii/S0169515097000042.

Tomberlin, D., & Holloway, G. (2008). Bayesian hierarchical estimation of technical efficiency in a fishery. *Applied Economics Letters, 17*(2), 201–204.

Ulimwengu, J. (2009). Farmers health and agricultural productivity in rural Ethiopia. *African Journal of Agricultural and Resource Economics, 3*(2), 83–100.

Ulimwengu, J. M. (2008). Farmers health and agricultural efficiency in rural Ethiopia: A stochastic production frontier approach.

USAID—The Ethiopian Government. (2011). President's malaria initiative: Ethiopia malaria operational plan *fy*2012. Technical Report, Ethiopia.

Wadud, A., & White, B. (2000). Farm household efficiency in Bangladesh: A comparison of stochastic frontier and DEA methods. *Applied Economics, 32*(13), 1665–1673.

Wang, H., & Schmidt, P. (2002). One-step and two-step estimation of the effects of exogenous variables on technical efficiency levels. *Journal of Productivity Analysis, 18*(2), 129–144.

Welsch, D. E. (1965). Response to economic incentive by Abakaliki rice farmers in eastern Nigeria. *Journal of Farm Economics, 47*(4), 900–914. http://www.jstor.org/stable/1236333.

W.H.O. (2007). Insecticide -treated mosquito nets: A W.H.O position statement. Technical report.

W.H.O. (2016). Ethiopia—malaria. http://www.afro.who.int/en/Ethiopia/country-programmes/topics/4580-Ethiopia-malaria.html.

Yotopoulos, P. A., Lau, L. J., & Lin, W.-L. (1976). Microeconomic output supply and factor demand functions in the agriculture of the province of Taiwan. *American Journal of Agricultural Economics, 58*(2), 333–340.

Zieschang, K. D. (1983). A note on the decomposition of cost efficiency into technical and allocative components. *Journal of Econometrics, 23*(3), 401–405.

Alternative Distribution Channels of Fruits and Vegetables

Kallirroi Nikolaou, Efthimia Tsakiridou, Foivos Anastasiadis and Konstadinos Mattas

Abstract The current study aims to explore consumer attitudes and preferences towards alternative food distribution channels focusing mainly on fresh fruits and vegetables. Consumer behavior towards alternative channels and various factors affecting purchasing decision are investigated and recorded, demonstrating the value of alternative channels in distributing agro-food products. Results manifest alternative channels are trusted by the consumers and they can be expanded particularly for fresh fruits and vegetables.

Keywords Sustainable supply chain management · Alternative distribution channels · Fruits and vegetables · Consumer behavior · E-commerce

The ever-increasing demand for food caused a huge effort to increase food production via the introduction of various forms of technologies. Specifically, technologies such as satellite navigation, sensor network, grid computing, ubiquitous computing, and context-aware computing are being applied more and more in the field, which help improve monitoring and increase decision-making capabilities (Rehman et al. 2014). However, the adoption of innovative technologies expanded not only to the production side but also to the distribution side (Wen 2007). Therefore, examining consumers' acceptance of technology will facilitate development of more efficient food distribution channels. The next section offers a short review of the relevant literature followed by the methodology description and results. Finally, the paper ends with the main conclusions and implications.

K. Nikolaou · E. Tsakiridou (✉) · K. Mattas
Department of Agricultural Economics, School of Agriculture,
Aristotle University of Thessaloniki, Thessaloniki, Greece
e-mail: efitsaki@auth.gr

F. Anastasiadis
Industrial Management Division of the Mechanical Engineering Department,
School of Engineering, Aristotle University of Thessaloniki, Thessaloniki, Greece

© Springer International Publishing AG, part of Springer Nature 2018
K. Mattas et al. (eds.), *Sustainable Agriculture and Food Security*,
Cooperative Management, https://doi.org/10.1007/978-3-319-77122-9_8

1 Literature Review and Theoretical Background

How and what an individual decides to buy is based on complex personal decisions. One of the factors affecting a consumer's purchasing behavior is his/her personality (Kassarjian 1971; Chen et al. 2015). Additionally, studies have shown that consumer behaviour is influenced by a person's socio-economic status, the environment, personal relationships, price, firm and also packaging (Asp 1999; Peter and Olson 2008; Yagi 2014). There is also a scientific interest towards investigating the measurement of consumer ecological behavior, environmental knowledge, healthy food, and a healthy way of life (Norazah 2014). Another critical factor is the role of technology. Sanyang et al. (2009) examine technology development and transfer to come up with appropriate strategies in the field of production and marketing of vegetables.

There has been a substantial amount of research into consumers' purchasing behavior over the last decades. Studies have examined more specifically their attitudes and behavior to environmentally friendly products (Davis 1992; Ohtomo and Hirose 2007) revealing that the five values that affect consumers' perceptions and subsequently influence their sustainable purchasing behaviour of green products are the following: (1) functional value, which involves price and quality; (2) epistemic value, which is the need to have knowledge; (3) social value, which has to do with image concerns and peer opinion; (4) conditional value, which is how promotional campaigns, government subsidies, etc. influence consumers; and (5) environmental value, which is the need or desire to take an active role towards protecting the environment (Laroche et al. 2001; Sharma and Bagoria 2012).

Research on environmentally friendly products and consumer perceptions, attitudes, apprehension and behavior has been conducted in Europe and America (Saxena and Khandelwal 2010; Biswas and Roy 2015), with a lesser number in the rapidly growing economies of the East (e.g. India) (Swaminathan et al. 1999). A number of studies have found that there are particular factors that affect consumer attitudes and behavior regarding the purchasing of fresh fruit and vegetables (Wong et al. 1996; Fisk 1998; Nair and Menon 2008; Moser et al. 2011; Fuentes 2014); in particular there are internal and external factors that influence consumer behaviour, like the individual's personal needs, motivation and perceptions, and on the other hand, the socio-economic and cultural environment. The global economic crisis is an external factor that has had a significant impact on consumer behaviour (Solomon 2009). The international literature shows that consumers tend to be favourably disposed to the purchase of environmentally friendly products and show interest in more sustainable ways of production even in a period of economic instability (Rehman et al. 2014; Radman 2005; Lea and Worsley 2008). The study by Gil et al. (2001), and also a more recent one by Ohtomo and Hirose (2007), showed that environmental characteristics are more important for fresh and perishable products.

Rödiger and Hamm (2015) investigated organic food and how consumer behavior was affected by price fruits and vegetables. Carlucci et al. (2015) studied consumers' purchasing behavior towards fish and seafood products and provided some patterns and insights from a sample of international studies. More specifically, this study presented a systematic review to assess consumer purchasing behavior towards fish and seafood products in developed countries.

There has been growing interest worldwide regarding product sustainability. Carayannis and Rakhmafullin (2014) explore and explain the value of smart specialisation strategies for sustainable and inclusive growth in Europe. Grunert et al. (2014) investigate the relationship between consumer incentive, perception and the application of sustainability labels on food products. The findings show that there were significant levels of concern with sustainability issues and that consumer motivation is an important factor. Carvalho et al. (2015) applied a 19-item scale to measure the behavioral factors that motivate a consumer to purchase products and services that are environmentally sustainable.

Consumer behavior in purchasing minimally processed vegetables and the implications for marketing strategies have been studied by Sillani and Nassivera (2015). The influence of packaging materials on consumer choice has been investigated as well. Food-packaging life cycle assessment studies showed that packaging systems should take into account food waste, which has significant environmental impacts (Wikström et al. 2014). The growing consumer desire for fresh, tasty food with reduced packaging and little or no processing and the increase of consumers' demand for a wide variety of fresh fruit and vegetables drive the research towards studying the fresh fruit and vegetable market. In addition, the influence of social media on consumer behavior, combined with the above, offers new insights in consumer behavior. They assessed Twitter food-related data. They argue that the increase of references to eating and drinking on Twitter, one of the most popular social media, appears to have potential in influencing consumer behavior. The following two studies examine technology applications for agriculture. The study of Long et al. (2013) could also be mentioned here on the effectiveness of cell phones and mypyramidtracker.gov on fruit and vegetable intake.

Research has also shown that consumers are concerned with processes and production techniques and technologies (Chinnici et al. 2002; Dimara and Skuras 2005). Furthermore, there are many factors that influence the technology transfer (Zhao and Grier 1991). The key role of technology transfer is revealed from the research of Audretsch et al. (2016). Thus, there are many studies on the improvement of flow distribution, showing the impact of technology in the field of distribution channels (Pistoresi et al. 2015). Watson IV et al. (2015) have provided a comprehensive review of marketing channels' research from 1980 to 2014, stating that the field of distribution channels is wide. It is precisely this span that enables us to incorporate the subject into our study.

Based on the literature review the main concept revealed was 'e-commerce'; that initially sprung from the need of some consumers who were not able to shop at physical stores. This gradually evolved to encompass a wider part of the society, largely due to the benefits of online shopping, which include: convenience, time-saving, fewer 'luring' traps, variety and better prices. Having to make a purchasing decision, however, involves risk (i.e. due to the online shopping environment) and may provoke feelings of confusion and insecurity to consumers. Their attitude to trust, known as their 'trust propensity' in these types of new shopping experiences, plays a big role in determining what action they will take, (Gefen 2000; McKnight and Chervany 2001) and whether or not they respond positively to the information they are given by others (Gefen et al. 2008). In addition, a growing number of researchers have been involved in developing knowledge-based information systems that help managers in the online decision-making process generally (Matsatsinis and Siskos 1999; Özbayrak and Bell 2003), while Wen (2007) has built an "intelligent knowledge-based e-commerce system" specifically for the purchase of agricultural products. Furthermore, agricultural products are very sensitive and spoilage may occur. A long distribution chain may damage the freshness of fruits and vegetables. An effective e-commerce system may be a necessary innovation to help farmers sell their agricultural products online (Wen 2007).

Technological changes occurred over the last few years, transforming agrifood systems into much more complex ones (Lowe et al. 2008). As a result of the lack of human contact, a distribution channel of agricultural products like e-commerce cannot have one-to-one interaction, making the entire process not only impersonal but also of limited scope due to the perishable nature of the products being provided. Thus, the online purchase of fresh fruit and vegetables must occur on or close to the same day and locally or at most regionally. On the whole, despite the fact that in order to shop online, one must not only possess the necessary technology but also be e-literate. As a form of alternative distribution, e-commerce has very quickly become an accepted means of product distribution for both providers and consumers (Wen 2007).

Furthermore, as socio-economic characteristics have an influence on consumer behavior, the correlations between these two variables are investigated. Generally, 'e-commerce' as a distribution channel has grown to facilitate the agrifood sector.

The above-mentioned studies reveal an increased interest in investigating consumer attitudes and behavior towards perishable products and the effectiveness of the channels. Thus, the current paper aims at contributing to the field of technology application in the channels of fruit and vegetables.

2 Methodology

A variety of techniques and approaches have been proposed and applied in measuring attitudes (Maison et al. 2004; Malhotra 2007; Jiao et al. 2012; Gesinde et al. 2014). The employed methodology helps to explore how people face a particular issue and how population characteristics, which include qualities and characterization of various types of populations, health status, socioeconomic factors, and consumer needs are interrelated, affecting buying patterns. Based on the above, the following hypotheses have been developed:

H1: There is a relationship between the use of technology for purchasing agri-food products and consumers' socio-economic characteristics.

H2: There is a relationship between purchasing agrifood products from alternative distribution channels and consumers' socio-economic characteristics.

The investigation of the two hypotheses was conducted via Hypothesis Testing, using a statistical test to determine whether there is enough evidence in a sample of data to infer that a certain condition is true for the entire population. In similar studies on consumer behavior with normal distribution, hypothesis testing has also been applied (Almossawi 2014; Bilgin and Brenner 2013; Siegel 2012; Smith 2012).

In order to test H1, the control variables are 'e-commerce' and 'technology application in a distribution channel' and the socio-economic factors 'gender', 'age' and 'level of education'. Furthermore, in order to test H2 the control variables are in the first step all types of distribution channels with socio-economic factors. However, due to the focus on alternative distribution channels, in the second step, we exclude the main distribution channel, especially the variables 'farmers' market', 'grocery' and 'supermarket' which constitute the core of main distribution channels.

3 Results and Discussion

Profile of the participants: Among four hundred consumers, 57% were male and 43% female. In addition, 75% had a university degree and 29.75% were graduates of technical education institutes. 24.5% were aged 18–36 years, 23.25% were aged 36–45 years, 32.5% were aged 46–55 years and 19.75% were aged 56–65 years. Furthermore, income range starts from 1001 to 1500€ (27.75%), up to 2001–2500 € (22.75%) and the rest were in between (32%). The tested hypothesis shows the correlations between the eight (8) pairs of variables: (i) 'e-commerce' and 'age'. (ii) 'e-commerce' and 'level of education', (iii) 'level of education' and 'technology application in a distribution channel', (iv) 'monthly family income' and 'technology application in a distribution channel', (v) 'level of education' and 'directly from producer', (vi) 'level of education' and 'organic market', (vii) 'e-commerce' and 'monthly family income', and (viii) 'level of education' and 'non-intermediate traders of fruits and vegetables' (presented in Tables 1 and 2). Firstly, Table 1 shows that four pairs of random variables were statistically significant regarding factors that influence consumer behavior between socio-economic factors, resulting in the acceptance of H1. More specifically, Table 1 shows the relationship between the variables use of technology for purchasing agrifood products and consumers' socio-economic characteristics. Socio-economic characteristics also influence sustainable consumption of fresh fruits and vegetables.

More specifically, the socio-economic factor 'age' showed a significant correlation with 'e-commerce' (Asymp. Sig: 0.017). In addition, the socio-economic factor 'level of education' had a significant correlation with the following variables:

Table 1 Hypothesis testing influence between the use of technology for purchasing agrifood products and consumers' socio-economic characteristics

Variables		Asymp. Sig
'age'	'e-commerce'	0.017
'level of education'	'e-commerce'	0.000
'level of education'	'technology application in a distribution channel'	0.016
'monthly family income'	'technology application in a distribution channel'	0.000

Table 2 Hypothesis testing of socio-economic factors towards alternative distribution channels of fruits and vegetables

Variables		Asymp. Sig
'directly from the producer'	'level of education'	0.001
'Organic Market'	'level of education'	0.003
'e-commerce'	'monthly family income'	0.002
'non-intermediate traders of fruits and vegetables'	'level of education'	0.036

'e-commerce' (Asymp. Sig: 0.000) and 'technology application in a distribution channel' (Asymp. Sig: 0.016). Lastly, the variable 'monthly family income' in conjunction with 'technology application in a distribution channel' had a significance of Asymp. Sig: 0.000. Thus, the findings from Table 1 clearly show that the most significant variables that influence consumers to use e-commerce for fruits and vegetables are level of education, monthly income, and age. E-commerce enables consumers to shop on-line 24 h per day, 365 days per year. The time-consuming parts of shopping at physical stores, driving, parking and coping with traffic are thus all eliminated. This factor is raising optimism in e-commerce, which rests on the hope that people will spend more online than they would in conventional shopping behavior. More particularly, the relation between technology and socioeconomic factors clearly emerged from this study. It was found that the younger a consumer is, the more likely they are to use e-commerce. This would then indicate that there is room for improvement in training older consumers to use the internet and by extension e-commerce, especially where fresh fruit and vegetables are concerned. This is where the contribution of the technological revolution in agriculture in conjunction with socioeconomic factors is revealed. In areas of low education level and monthly income, the opportunity is given to both traders and the state to modernize the channels of distribution and improve these levels of social factors. These specific factors, namely level of education and monthly income, include/add the dimension that the market for fresh fruit and vegetables comprises spirituality from a consumer's perspective. The higher the increase in these two factors, the more frequent is the use of technologies for the purchase of agricultural products.

Table 2 shows that four (4) pairs of random variables were statistically significant regarding socio-economic factors in relation to alternative distribution channels of fruits and vegetables, resulting in the acceptance of the H2 hypothesis.

The pair of variables '*directly from the producer*' and '*level of education*' had a significance of 0.001, and the variables 'organic market' and 'level of education' had a significance 0.003, meaning that higher educated people prefer to buy directly from the producer and from organic markets. Lastly, the pairs of variables: '*e-commerce*' and '*monthly family income*', '*non intermediate traders of fruits and vegetables*' and '*level of education*', had a significance of 0.002 and 0.036, respectively. This correlation shows that consumers with a high monthly family income use more e-commerce to buy fruits and vegetables, whereas highly educated consumers do not like many intermediates in the supply chain of fruits and vegetables. Summing up, in Table 2 again it can be seen that there is a significant correlation between the variable of alternative distribution channels of fruits and vegetables and the socio-economic factors of 'level of education', and 'monthly income' when it comes to consumer e-commerce support.

Furthermore, the variables 'level of education' and 'monthly family income' are related with the use of technology application in an agricultural product distribution. Therefore, the use of technology application in a distribution channel of fruits and vegetables is related to the education level and the monthly family income. In order to improve the use of e-commerce and the use of technology application, it is essential to improve the educational level and financial situation of the consumers.

The findings in Tables 1 and 2 show that the 'e-commerce' concept is acceptable as an alternative distribution channel of fruits and vegetables. This study investigates the correlation between socio-economic factors and agricultural product distribution channels could provide an integrated view on consumer behavior in alternative distribution channels including the key factor of technology via e-commerce. Overall, besides the impact of technology on consumer behaviour, the research findings can also be applied in determining the new trends and preferences, which if used to advantage by all interested parties, will stimulate purchases. The findings of this method of analysis overlap with the hypothesis testing results discussed above. Furthermore, this study presents potential for the development of the current situation concerning consumer behaviour in the purchase of fresh fruit and vegetables. This adds to the overall image of consumer behavior in distribution channel of fruits and vegetables. The tendency for consumers to support such a distribution channel indicates that it is crucial and probably vital for the sustainability of the enterprise to include e-commerce delivery. For example, if one trader uses e-commerce when he knows that consumers prefer this to the conventional channels, then he ensures the sustainability and the competiveness of his enterprise.

It is important to highlight some of the results and findings. One important result derived from this research is that age and e-commerce use are strongly related to the purchase of fresh fruits and vegetables (Table 1). Therefore, marketers can use this information to create accessible delivery methods of fresh fruit and vegetables that are targeted to particular age groups. Younger age groups are more familiar with the use of the internet, finding it easier to use e-commerce. For example, younger people who might not prefer fresh fruit and vegetables could become enticed to buy such products more often, as they would be readily available on an online application on their smart phones.

Consequently, e-commerce redefines the rules of doing business. In particular, this type of alternative distribution channel of fruits and vegetables could open the floodgates for the flow of local food by delivering food of superior taste, freshness, and nutritional value to many people. This is a very crucial implication that enhances the development of the agricultural sector, in general.

Lastly, e-commerce provides shoppers with new access to products and the ability to customize some items for their particular needs. Moreover, in an online environment, there are more opportunities to tailor pack and portion sizes towards different demographics. In particular, e-commerce provides a chance to producers to be prepared to supply more packaging formats.

4 Conclusion

It is a fact that developments in the last decades in agricultural markets have successfully adapted the flow of information technology to a wide range of cost-effective applications. Amid the diversity and change, patterns of e-commerce practices in agriculture are emerging. The limited research concerning sustainable agriculture in Greece was the main motivation for conducting this study. The findings on consumer behavior with respect to alternative and sustainable fruit and vegetables distribution channels contribute to the literature in the field of sustainable supply chain management. More specifically, the study findings reveal a number of significant characteristics that define the sustainability of supply chain management.

Having a clear idea of consumer preferences greatly facilitates the development of effective policies and applications. The contribution of technological innovations, applied in agricultural distribution channels could develop a new environment and trends not only for consumers but also for entrepreneurs (producers—wholesalers—retailers). These applications ensure that fruits and vegetables will be delivered to consumers, whenever and wherever necessary. These applications give consumers the opportunity to discover the products of local shops and restaurants and taste their freshness. Through the use of these applications, consumers become accustomed to e-commerce and are more familiar with the way the distribution channels of fruits and vegetables work. More specifically, producers and/or individual entrepreneurs can take these findings into consideration so as to create useful promotion strategies. The present study's findings indicate that consumers support e-commerce as an alternative distribution channel of fruits and vegetables, and the purchase of fresh fruits and vegetables can further grow.

Furthermore, the present study provides directions to producers for altering their quality standards. The key implication of the present study is a valuable feedback to every stakeholder involved in an alternative supply chain. Such feedback could bring together Science and Business: producers can become acquainted first-hand with customer attitudes and needs—knowledge which allows them to make innovative, profitable choices that are both timely and cost-effective—while managers can build upon current trends with a more efficient decision-making process.

Lastly, the study findings contribute to society as a whole and can be applied to bring together and engage consumers and producers. Particularly, producers can use the findings to form a clearer picture of what customers actually want and need from the product purchase of fruit and vegetables. Then, producers will gain the knowledge allowing them to make innovative, effective entrepreneurial decisions. Fruit and vegetable policy makers could benefit as well, reforming current legislation and adopting more environmentally friendly terms in modern distribution channels.

References

Almossawi, M. (2014). Promoting green purchase behavior to the youth. *British Journal of Marketing Studies, 2*, 1–16.

Asp, E. H. (1999). Factors affecting food decisions made by individual consumers. *Food Policy, 24*, 287–294.

Audretsch, D., Lehmann, E., Paleari, S., & Vismara, S. (2016). Entrepreneurial finance and technology transfer. *The Journal of Technology Transfer, 41*, 1–9.

Bilgin, B., & Brenner, L. (2013). Context affects the interpretation of low but not high numerical probabilities: A hypothesis testing account of subjective probability. *Organizational Behavior and Human Decision Processes, 121*, 118–128.

Biswas, A., & Roy, M. (2015). Green products: An exploratory study on the consumer behavior in emerging economies of the East. *Journal of Cleaner Production, 87*, 463–468.

Carayannis, E. G., & Rakhmatullin, R. (2014). The quadruple/quintuple innovation helixes and smart specialization strategies for sustainable and inclusive growth in Europe and Beyond. *Journal of the Knowledge Economy, 5*, 212–239.

Carlucci, D., Nocella, G., Devitiis, B., Viscecchia, R., Bimbo, F., & Nardone, G. (2015). Consumer purchasing behaviour towards fish and seafood products. Patterns and insights from a sample of international studies. *Appetite, 84*, 212–227.

Carvalho, B. L., Salgueiro, D. F., & Rita, P. (2015). Consumer Sustainability Consciousness: A five dimensional construct. *Ecological Indicators, 58*, 402–410.

Chen, Y., Yan, X., Fan, W., & Gordon, M. (2015). The joint moderating role of trust propensity and gender on consumers' online shopping behavior. *Computers in Human Behavior, 43*, 272–283.

Chinnici, G., Amico, M., & Pecorino, B. (2002). A multivariate statistical analysis on the consumers of organic products. *British Food Journal, 104*, 187–199.

Davis, J. (1992). Ethics and environmental marketing. *Journal of Business Ethics, 11*, 81–87.

Dimara, E., & Skuras, D. (2005). Consumers' demand for informative labeling of quality food and drink products: A European union case study. *Journal of Consumer Marketing, 22*, 90–100.

Fisk, G. (1998). Green marketing: Multiplier for appropriate technology transfer? *Journal of Marketing Management, 14*, 657–676.

Fuentes, C. (2014). How green marketing works: Practices, materialities, and images. *Scandinavian Journal of Management, 31*, 192–205.

Gefen, D. (2000). E-commerce: The role of familiarity and trust. *Omega, 28*, 725–737.

Gefen, D., Benbasat, I., & Pavlou, P. (2008). A research agenda for trust in online environments. *Journal of Management Information Systems, 24*, 275–286.

Gesinde, A. M., Temitope, A., & David, O. I. (2014). The development and validation of vignette-type instrument for measuring attitude toward poverty. *Procedia—Social and Behavioral Sciences, 159*, 442–446.

Gil, N., Tommelein, I. D., & Kirkendall, R. L. (2001). Modeling design development in unpredictable environments. In B. A. Peters, J. S. Smith, O. J. Medeiros & M. W. Rohrer (Eds.), *Proccedings of Winter Simulation Conference (WSC)* (pp. 515–522), December 01–09, 2001, Arlington VA.

Grunert, K. G., Hieke, S., & Wills, J. (2014). Sustainability labels on food products: Consumer motivation, understanding and use. *Food Policy, 44*, 177–189.

Jiao, Y., Zhou, H., Wang, J., & LI, J. (2012). Linearization error's measure and its influence on the accuracy of MEKF based attitude determination method. *Aerospace Science and Technology, 16*, 61–69.

Kassarjian, H. (1971). Personality and consumer behavior: A review. *Journal of Marketing Research, 8*, 409–418.

Kozlenkova, I. V., Hult, T. M., Lund, D. J., Mena, J. A., & Kekec, P. (2015). The role of marketing channels in supply chain management. *Journal of Retailing* (in press).

Laroche, M., Bergeron, J., & Barbaro-Forleo, G. (2001). Targeting consumers who are willing to pay more for environmentally friendly products. *Journal of Consumer Marketing, 18*, 503–520.

Lea, E., & Worsley, A. (2008). Australian consumers' food-related environmental beliefs and behaviours. *Appetite, 50*, 207–214.

Long, J. D., Boswell, C., Rogers, T. J., Littlefield, L. A., Estep, G., Shriver, B. J., et al. (2013). Effectiveness of cell phones and mypyramidtracker.gov to estimate fruit and vegetable intake. *Applied Nursing Research, 26*, 17–23.

Lowe, P., Phillipson, J., & Lee, R. P. (2008). Socio-technical innovation for sustainable food chains: roles for social science. *Trends in Food Science & Technology, 19*, 226–233.

Maison, D., Greenwald, A. G., & Bruin, R. H. (2004). Predictive validy of the implicit association test in studies of brands, consumer attitudes, and behavior. *Journal of Consumer Psychology, 14*, 405–415.

Malhotra, N. K. (2007). *Marketing research: An applied orientation.* New Jersey.

Matsatsinis, N. F., & Siskos, Y. (1999). MARKEX: An intelligent decision support system for product development decisions. *European Journal of Operational Research, 113*, 336–354.

Mcknight, D. H., & Chervany, N. L. (2001). Trust and distrust definitions: One bite at a time. *In Trust in Cyber-societies, LNAI, 2246*, 27–54.

Moser, R., Raffaelli, R., & Thilmany-McFadden, D. (2011). Consumer preferences for fruit and vegetables with credence-based attributes: A review. *International Food and Agribusiness Management Review, 14*, 121–142.

Nair, S. R., & Menon, C. G. (2008). An environmental marketing system—a proposed model based on Indian experience. *Business Strategy and the Environment, 17*, 467–479.

Norazah, M. S. (2014). Investigating the measurement of consumer ecological behavior, environmental knowledge, healthy food, and healthy way of life. *Social Ecology and Sustainable Development, 5*, 1–10.

Ohtomo, S., & Hirose, Y. (2007). The dual-process of reactive and intentional decision-making involved in eco-friendly behavior. *Journal of Environmental Psychology, 27*, 117–125.

Özbayrak, M., & Bell, R. (2003). A knowledge-based decision support system for the management of parts and tools in FMS. *Decision Support Systems, 35*, 487–515.

Peter, J. P., & Olson, J. C. (2008). *Consumer behaviour and marketing strategy* (7th ed.). Mcgraw-Hill/Irwin Series in Marketing.

Pistoresi, C., Fan, Y., & Luo, L. (2015). Numerical study on the improvement of flow distribution uniformity among parallel mini-channels. *Chemical Engineering and Processing: Process Intensification, 95*, 63–71.

Radman, M. (2005). Consumer consuption and perception of organic products in Croatia. *British Food Journal, 107*, 263–273.

Rehman, A., Abbasi, Z., Islam, N., & Shaikh, Z. (2014). A review of wireless sensors and networks' applications in agriculture. *Computer Standards & Interfaces, 36*, 263–270.

Rödiger, M., & Hamm, U. (2015). How are organic food prices affecting consumer behaviour? A review. *Food Quality and Preference, 43*, 10–20.

Sanyang, S., Kao.,T., & Haung, W. (2009). Comparative study of sustainable and non-sustainable interventions in technology development and transfer to the women's vegetable gardens in the Gambia. *The Journal of Technology Transfer, 34*, 59–75.

Saxena, R., & Khandelwal, P. (2010). Can green marketing be used as a tool for sustainable growth? A study performed on consumers in India—An emerging economy. *The International Journal of Environmental, Cultural, Economic and Social Sustainability, 2*, 277–291.

Sharma, S. C., & Bagoria, H. (2012). Green marketing: A gimmick or the real deal? *International Journal of Research in Finance Marketing, 2*, 406–414.

Siegel, A. F. (2012). Chapter 10—Hypothesis testing: Deciding between reality and coincidence. In A. F. Siegel (Ed.), *Practical business statistics* (6th ed.). Boston: Academic Press.

Sillani, S., & Nassivera, F. (2015). Consumer behavior in choice of minimally processed vegetables and implications for marketing strategies. *Trends in Food Science & Technology* (in press).

Smith, G. 2012. Chapter 7—Hypothesis testing. In G. Smith (Ed.), *Essential statistics, regression, and econometrics*. Boston: Academic Press.

Solomon, M. (2009). *Consumer behavior, buying, having and being*. New Jersey, USA.

Swaminathan, V., Lepkowska-White, E., & RAO, B. P. (1999). Browsers or buyers in cyberspace? An investigation of factors influencing electronic exchange. *Journal of Computer-Mediated Communication, 5*, 1–18.

Watson IV, G. F., Worm, S., Palmatier, R. W., & Ganesan, S. (2015). The evolution of marketing channels: Trends and research directions. *Journal of Retailing* (in press).

Wen, W. (2007). A knowledge-based intelligent electronic commerce system for selling agricultural products. *Computers and Electronics in Agriculture, 57*, 33–46.

Wikström, F., Williams, H., Verghese, K., & Clune, S. (2014). The influence of packaging attributes on consumer behaviour in food-packaging life cycle assessment studies—a neglected topic. *Journal of Cleaner Production, 73*, 100–108.

Wong, V., Turner, W., & Stoneman, P. (1996). Marketing strategies and market prospects for environmentally-friendly consumer products 1. *British Journal of Management, 7*, 263–281.

Yagi, T. (2014). Knowledge creation by consumers and optimal strategies of firms. *Journal of the Knowledge Economy, 5*, 585–596.

Zhao, H., & Grier, D. (1991). Factors influencing technology transfer: The case of China. *The Journal of Technology Transfer, 16*, 50–56.

Kallirroi Nikolaou is Ph.D. candidate, in the Department of Agricultural Economics, from Aristotle University of Thessaloniki. Her education includes a B.Sc. in Agriculture from the Aristotle University of Thessaloniki (2011), an M.Sc. in Agricultural Economics from the Aristotle University of Thessaloniki (2012). Her research interests focus on alternative distribution channels of agricultural products.

Efthimia Tsakiridou is an Associate Professor at the Aristotle University of Thessaloniki, Greece. She has twelve years of research experience in the field of food marketing and supply chain management. She is a specialist in the field of agro-food marketing, supply chain management and consumer behavior and worked as a participant in several European and national research projects either as a participant or as a leading researcher. Dr. Tsakiridou has published in several refereed journals (Applied Economics, Journal of Food Products Marketing, Journal of International Food and Agribusiness Marketing, International Journal of Retail and Distribution Management, British Food Journal, Food Economics, International Journal of Economic Research).

Foivos Anastasiadis is a Postdoctoral Researcher at Aristotle University of Thessaloniki (A.U.Th.). He was specifically recruited as an expert on sustainable agrifood supply chains for the EU funded FP7 project GREEN-AgriChains. He holds an M.Sc. from Wageningen University, The Netherlands (2004) and a Ph.D. from Imperial College, London (2010). Prior to joining A.U.Th, Dr. Anastasiadis was employed: as an adjunct lecturer at the University of Macedonia, Greece; as an academic associate at the International Hellenic University, Greece; and as a consultant in the market research sector and in the supply chain/procurement processes business.

konstadinos mattas is a Professor of Agricultural Policy. His education includes a B.Sc. in Agriculture from the Aristotle University of Thessaloniki (1972), a B.Sc. in Economics from the University of Macedonia (1976), an M.Sc. in Agricultural Economics from the University of Kentucky (1982) and a Ph.D. in Agricultural Economics from the University of Kentucky (1984). He has published in more than 100 international refereed journals, in collective volumes and proceedings. He has contributed to international and Greek conferences and served as a referee to several international journals.

Evaluation of Irrigation Efficiency Effect on Groundwater Level Variation by Modflow and Weap Models: A Case Study from Tuyserkan Plain, Hamedan, Iran

Abdollah Taheri Tizro, Konstantinos Voudouris, Christos Mattas, Morteza Kamali and Meysam Rabanifar

Abstract Concerning the increasing demand for water in arid and semiarid regions, especially in the field of agriculture, an analysis of irrigation efficiency plays a vital role. One of the most important approaches for increasing efficiency in agriculture is to develop modern systems and use them to replace traditional irrigation methods. Tuyserkan Plain is located in Hamadan province, Iran, that is faced with a severe decline of groundwater level. Irregular use of groundwater for agricultural needs and low-efficiency irrigation systems are believed to be the main reasons. In this study, the efficiency of traditional and modern irrigation systems is first analyzed. An irrigation efficiency scenario is presented in order to simulate irrigation efficiency increase effects on the level of groundwater in the aquifer. The irrigation efficiency scenario is simulated using the MODFLOW and WEAP models and is compared to the reference scenario (Current irrigation efficiency is maintained). Although groundwater level declines in both scenarios, the simulation showed that in the reference scenario, groundwater level decrease becomes significantly slower and, as a result, the aquifer will recover.

Keywords Irrigation efficiency · Tuyserkan plain · Groundwater
MODFLOW · WEAP · Simulation

A. T. Tizro (✉)
Department of Water Engineering, College of Agriculture Bu-Ali Sina University,
Hamedan, Iran
e-mail: ttizro@yahoo.com

K. Voudouris · C. Mattas
Engineering Geology and Hydrogeology Lab, Department of Geology,
Aristotle University, Thessaloniki, Greece

M. Kamali · M. Rabanifar
College of Agriculture, Bu-Ali Sina University, Hamedan, Iran

© Springer International Publishing AG, part of Springer Nature 2018
K. Mattas et al. (eds.), *Sustainable Agriculture and Food Security*,
Cooperative Management, https://doi.org/10.1007/978-3-319-77122-9_9

1   Introduction

It is a vital necessity to maintain water resources in the face of increasing demands in agricultural, industrial and municipal uses. Production of crops and livestock is water-intensive; agriculture alone accounts for 70% of all water consumption (www.unesco.gr). The world will need 50% greater food supply for the population of 9.5 billion estimated for 2050 (Singh 2014). It is estimated that the water consumption for both rainfed and irrigated agriculture will increase by approximately 19% by 2050. About 30% of fresh water on earth is stored in aquifers (Shiklomanov and Rodda 2003). The amount of groundwater is more than 100 times greater than the amount of fresh surface water (in lakes and rivers) and its availability is not usually of seasonal character; therefore most of the cultivated lands in the world depend on groundwater. The recent years, many concerns have been raised regarding the climate change and its impacts on water resources. South Asia and Southern Africa are predicted to be the most affected regions due to climate change-related food shortages by 2030 (www.unesco.org). This has resulted in the lowering of the groundwater level (Pfeiffer and Lin 2014). Iran is considered to be one of the most water stressed countries in the world, since 73.8% of the total annual renewable fresh water is consumed (IMPO 2003; Jafary 2016). Iran is also one of the top countries in groundwater abstraction (Gleeson et al. 2012). Thus, it is rather imperative to improve and manage the drainage in catchments and necessary to understand the functions of irrigation. Moreover, the codification and application of laws concerning water usage in agriculture depend on the management of irrigation (Ahadi et al. 2013; Bournaris et al. 2015; Manos et al. 2009). Most policies seek to operate appropriate planning for irrigation or shift from irrigated to rainfed cultivations in order to reduce consumption. One of the most important methods is to develop more efficient irrigation systems (Cooley et al. 2009).

Models are among the most useful devices in water supply management, and especially multi-objective modelling tools that involve stakeholders and facilitate decision making. They can be utilized in gathering data and improving the process of planning and managing the water supply systems (Sulis and Sechi 2013; Taheri Tizro et al. 2011), combining physical and socioeconomic parameters. In the WEAP (Water Evaluation and Planning System) model (Stockholm Environment Institute 2005) standard linear programming is used for the water allocation problems. This allows the model to consider more complex physical, hydrological, and institutional constraints than the min-cost flow approach (Sulis and Sechi 2013). This model is a decision support system (DSS) and provides a complete analysis of water supply and demand at present and estimation for the process in the future (Yazdanpanah et al. 2008). This software is used when there is multipurpose and competitive demand and allows the analysis of different management patterns (Seiber et al. 2005). The MODFLOW-3D finite difference groundwater flow model is able to simulate groundwater flow and different complex hydrologic situations by a numerical system. Its flexibility, complete coverage of hydrological process and availability has made it a popular device at a global scale (Zhou and Li 2011).

Zadevakili (2011) used both WEAP and MODFLOW models to study the possibility of a combined utilization of both surface and groundwater in Zayanderood Basin. They studied the effects of different management policies on industrial, agricultural and municipal demands from the basin and the sustainability of the aquifer. The simulation results showed that in order to utilize the basin as a reliable long-term resource and a supplement for surface water, and also in order to improve the condition of the aquifer, several important factors are to be considered. These factors include changing the consumption pattern thus decreasing demands, correcting aquifer discharge, transferring water to the basin and altering distribution priorities.

Mehta et al. (2013) studied the potential effect of climate changes on Cache Creek watershed in California until 2099 by conducting three different scenarios in the WEAP model:

(1) Changes based on economical estimations.
(2) Utilizing more diverse crop patterns with less water demand.
(3) Combining different irrigation technologies with changing crop patterns in order to provide crops with less water demand.

The study pointed out that the third scenario will be more successful in decreasing water demand and will lower the water demand by 12%, thus also decreasing groundwater discharges.

Alizadeh et al. (2014) studied the effects of developing pressurized irrigation systems on the Varamin Plain. They analyzed the issue by simulating in three scenarios:

(1) Current circumstances.
(2) Developing pressurized irrigation systems by 10% without increasing the cultivation area.
(3) Increasing pressurized irrigation systems by 10% by increasing the cultivation area.

The results showed that developing pressurized irrigation systems by 10% either with increasing the cultivation level by 20% or without it will improve the quantity of groundwater.

Mahdavi et al. (2012) studied the management operations on the aquifer in the Hamedan-Bahar Plain by using a MODFLOW model. They analyzed and estimated the aquifer's condition in a five-year period in different situations including: current discharge without an increase in rainfall, current discharge with a 20% increase in rainfall average, changing crop pattern, changing irrigation methods and shutting down illegal wells. These studies estimated that in five years, the current amount of discharges will result in lowering the water level, regardless of the rainfall increase, but the other three options will improve the aquifer's condition. Changing the irrigation method will have the highest effect compared to other options.

This study initially analyses different crop patterns, the amount of groundwater discharge and different irrigation systems in the Tuyserkan plain. Then, irrigation efficiency will be determined. In the next step, after gathering sufficient data, WEAP and MODFLOW models are used to simulate the aquifer condition and supply-demand modeling. Finally, the effects of increasing irrigation efficiency on groundwater are analyzed.

2 Materials and Methods

Tuyserkan Plain

Tuyserkan Plain (805 km^2) is one of the plains in the upper Karkheh river basin and is located to the south of Mount Alvand, Hamadan province, Iran. The plain is positioned between the longitudes of 48° 05′–48° 35′ E and latitudes of 34° 22′–34° 42′ N. The aquifer has an area of about 150 Km^2. The plains' geographical situation and topography is shown in Fig. 1.

Based on the data provided from Hamadan's Jihad of Agriculture Organization, the whole of Tuyserkan County had 38274 hectares of cultivated land in the 2010–2011 farming year. Approximately 8890 hectares from Tuyserkan were in irrigated farming form and 5405 hectares consisted of food trees. Overall, 14295 hectares were utilised for irrigated agriculture. Due to lack of sufficient data on crop water consumptive use in individual villages, the data were assorted for each rural district. There are four rural districts in the Toyserkan area, including: Ghogholrood, Hayaghoogh Nabi, Korzanrood and parts of Mianrood. These districts and the installed piezometers are shown in Fig. 2.

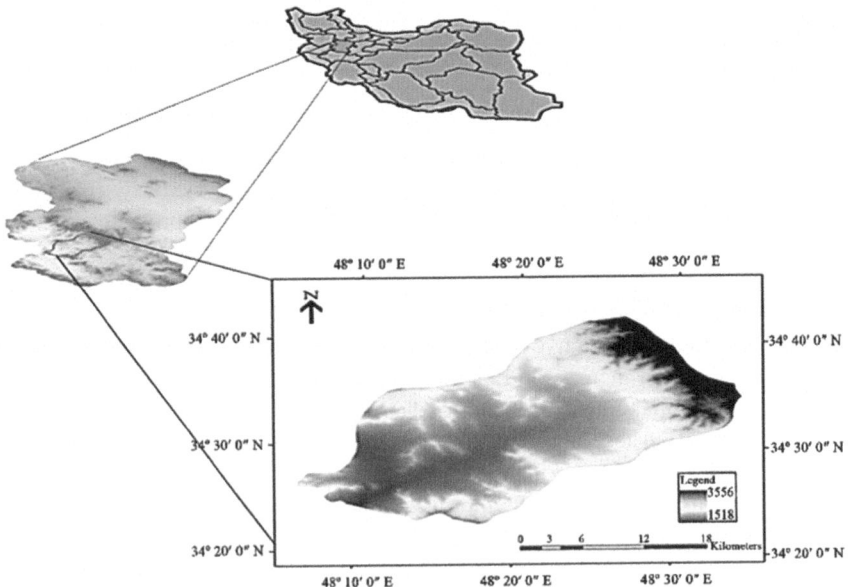

Fig. 1 Tuyserkan plain in Hamadan province Hamadan's Jihad of agriculture organization

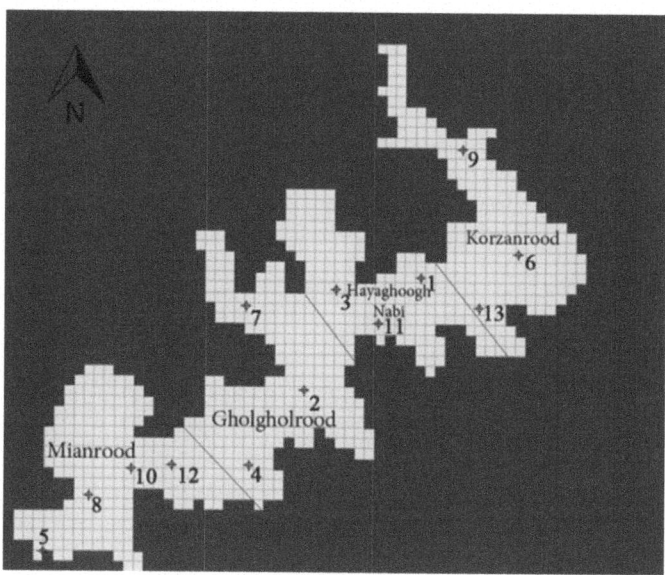

Fig. 2 Piezometers position with agricultural sites in Tuyserkan aquifer

Based on Hamadan's Jihad of Agriculture Organization, crop pattern and cultivated area are respectively shown in Tables 1 and 2. Like many other regions in Iran, cereals are the most produced crops in the Tuyserkan region (46.6% of the total and about 70% of crop products). Food trees cover 5405.5 hectares in this region. Walnut tree is the main food crops in the region (about 68.5% of orchard products and 23% of the overall agricultural crop). According to data provided by Hamadan's Jihad of Agriculture Organization and the Regional Water Organization of Hamadan (2008), crop water consumption in the Tuyserkan Plain is about 122.45 Mm^3 from groundwater resources.

It is necessary to determine the total irrigation efficiency for both traditional and pressurized irrigation methods in the region, in order to realize the amount of water demanded by each type of cultivation. Since there has been no assessment of irrigation methods in the Tuyserkan Plain, total irrigation efficiency was measured by crop water consumption data provided from Hamadan Province's Spatial Planning and also statistics from Hamadan's Jihad of Agriculture Organization.

Analysis of pressurized irrigation

In the Tuyserkan Plain, 3311.47 hectares of land, use the pressurized irrigation system. Different types of these systems include the classic sprinkler, gun and drip system. Almost all of this area in the Tuyserkan Plain uses groundwater resources. By considering different factors such as water discharge, cultivated land covered by wells and types of cultivation, the efficiency of each system was analyzed. Irrigation

Table 1 Irrigated farming pattern and cultivated area of them in agricultural sites in Tuyserkan plain (*Source* Regional Water Organization of Hamadan 2008)

Crop	Cultivated area in Mianrood site (hectares)	Cultivated area in Korzanrood site (hectares)	Cultivated area in Gholgholrood site (hectares)	Cultivated area in Hayaghoogh Nabi site (hectares)	Total of cultivated area in Tuyserkan plain (hectares)
Wheat	1036.21	168	1307	1585.4	4096.61
Oats	216.62	105	401	506.5	1229.1
Sugar beet	133	0	168	0	300.96
Potato	7.25	3.6	25	179.45	161.45
Alfalfa	170.20	58	411	184.7	823.89
Clover	107.20	20	112	83.7	323.52
Garlic	21.5	15.5	25	292.8	354.8
Tomato	49.32	3.5	65	45.7	163.5
Cucumber	164.88	0	147	23.3	335.17
Corn	120.88	0	108	0	228.8
Helianthus	80.75	0	2	43.0	385.39
Chickpea	92.11	1	169	8.1	10.55
Bean	48.86	12	65	166.65	292.48
Vegetables	48.22	0	17	5.2	70.42

Table 2 Cultivated area of food trees in agricultural sites in Tuyserkan plain (*Source* Regional Water Organization of Hamadan 2008)

Crop	Cultivated area in Mianrood site (hectares)	Cultivated area in Korzanrood site (hectares)	Cultivated area in Gholgholrood site (hectares)	Cultivated area in Hayaghoogh Nabi site (hectares)	Total of cultivated area in Tuyserkan plain (hectares)
Walnut	128.46	693	251	2301	3373.46
Grapes	76.34	0	155	127.8	359.14
Almonds	6.52	23.5	66	111.5	207.52
Apple	27.51	11	166	101.8	306.3
Cherry	6.71	12.6	13	68.7	101.01
Plum	27.73	28	30	62.4	144.13
Pear	2.75	6	5	25.6	39.35
Peach	7.24	56.5	115	77.8	256.54
Others	1.97	48.4	6	128.1	184.47

efficiency is defined (Eq. 1) as the amount of effective water in relation to total water usage (Burt et al. 1997).

$$\text{Irrigation efficiency} = \frac{\text{Effective water}}{\text{Total water usage}} \tag{1}$$

If groundwater is utilized to provide for water usage, we can use Eq. (2) to determine irrigation efficiency.

$$\text{Total irrigation efficiency} = \frac{\text{Net requirement (m}^3 \text{ per year)}}{\text{Amount of water discharge (m}^3 \text{ per year)}} \tag{2}$$

Irrigation systems with higher efficiency increase this ratio, thus resulting in a lowering of the amount of water usage for a particular crop. In this study, Eq. (2) was used. Data of each site were provided by Hamadan's Jihad of Agriculture Organization and the Regional Water Organization of Hamadan. 11 samples from drip irrigation, 21 samples from gun and 13 samples from permanent and semi-portable classic sprinkler irrigation were used. Net requirement irrigation for farming and food trees in Tuyserkan Plain is shown in Tables 3 and 4, respectively. Table 5 shows the results of each pressurized system's efficiency.

Analysis of surface irrigation
The efficiency of surface irrigation was also determined by considering different factors such as water discharge, cultivated land covered by wells and types of cultivation. The efficiency of each method was analyzed using Netwat software. Seven samples were used to calculate furrow irrigation efficiency and 51 samples for flood irrigation efficiency. Table 6 shows the results.

Table 3 Net requirement irrigation for irrigated farming in Tuyserkan plain (mm)

Crop	Jan	Feb	Mar	Apr	May	Jun	Jul	Aug	Sep	Oct	Nov	Dec	Total
Wheat	9	30	79	118	91	0	0	0	15	11	0	4	357
Oats	9	30	79	110	45	0	0	0	0	14	0	4	293
Sugar beet	0	0	0	26	120	181	172	112	21	0	0	0	632
Potato	0	0	0	31	106	173	169	117	0	0	0	0	596
Alfalfa	9	23	54	81	116	122	118	103	70	33	3	15	747
Clover	0	0	35	119	126	154	153	145	32	12	0	0	776
Garlic	9	30	79	115	80	0	0	0	10	12	0	4	339
Tomato	0	0	0	31	103	180	176	134	23	0	0	0	647
Cucumber	0	0	0	33	105	183	180	140	27	0	0	0	668
Corn	0	0	0	22	115	178	143	19	0	0	0	0	477
Helianthus	0	0	0	26	119	178	152	21	0	0	0	0	496
Chickpea	0	1	27	106	159	55	0	0	0	0	0	0	348
Bean	0	0	0	24	114	173	152	21	0	0	0	0	484
Vegetables	0	0	0	29	110	157	97	0	0	0	0	0	393

As displayed in Tables 5 and 6, replacing traditional irrigation with pressurized irrigation systems increases the total irrigation efficiency by up to 20%. By applying this increase in efficiency to agricultural lands, the comparison between the water requirement of crops and food trees in agricultural sites is shown in Tables 7 and 8.

Table 4 Net requirement irrigation for food trees in Tuyserkan plain (mm)

Crop	Jan	Feb	Mar	Apr	May	Jun	Jul	Aug	Sep	Oct	Nov	Dec	Total
Walnut	0	0	18	62	159	173	168	146	98	0	0	0	851
Grapes	0	0	0	25	125	134	126	48	0	0	0	0	458
Almonds	0	0	27	89	143	152	147	112	0	0	0	0	670
Apple	0	0	24	96	143	150	146	114	45	0	0	0	718
Cherry	0	0	0	74	143	150	140	98	0	0	0	0	605
Plum	0	0	0	70	135	143	138	113	43	0	0	0	642
Pear	0	0	24	96	143	150	146	114	45	0	0	0	718
Peach	0	0	16	78	135	143	138	119	64	0	0	0	693

Table 5 Total efficiency for pressurized systems in Tuyserkan plain

Irrigation system	Number of samples	Cultivated area (hectares)	Crop	Total efficiency
Permanent and semi-portable classic Sprinkler irrigation	13	130.36	Alfalfa, Wheat, Sugar beet and Potato	51.47
Gun	21	605.5	Alfalfa, Sugar beet, cereals	61.0
Drip irrigation	11	46.63	Apple, peach, walnut, grapes	56.36
The average efficiency of the entire systems	–	–	–	59.13

Table 6 Total efficiency for surface irrigation methods in Tuyserkan plain

Irrigation method	Number of samples	Cultivated area (hectares)	Crop	Cultivated area (hectares)	Total efficiency
Flooding irrigation	51	419.9	Alfalfa, Wheat, Oats, Corn, Bean, peach, walnut, almonds	419.9	41.38
Furrow irrigation	7	150.8	Alfalfa, Sugar beet, cereals	150.8	34.77
The average efficiency of the entire methods	–	–	–	–	39.63

Table 7 Annual water requirement for crops and food trees in the absence of mechanized irrigation (m³)

Crop	Hayaghoogh Nabi	Gholgholrood	Korzanrood	Mianrood	Tuyserkan plain
Crops	19640267	28054933	4173798	22872761	74741759
Food trees	32727662	5216282	11746209	4947965	54638117
Total	52367929	33271215	15920006	27820725	129379876

Table 8 Annual water requirement for crops and food trees if all lands equipment with mechanized irrigation (m³)

Crop	Hayaghoogh Nabi	Gholgholrood	Korzanrood	Mianrood	Tuyserkan plain
Crops	13163263.99	18802925.76	2797354.98	15329739.73	50093284.46
Food trees	21934673.39	3496046.84	7872522.4	3314912.93	36618155.57
Total	35097937.38	22298972.6	10669877.39	18644652.66	86711440.03

WEAP model

In the WEAP model, a supply and demand simulation is used to determine the amount of distributable water for consumption areas by considering the amount of allocation water resources. Data from 2011 were interred to the model as the reference year. The data includes municipal, agricultural and industrial water usage in addition to return flow. The water resource in this study is groundwater. The parameters used for groundwater are storage capacity, initial storage and natural recharge. The data from reference year (2011) were used to calibrate the model. Groundwater volume from WEAP model is converted to saturated depth by storage coefficient and aquifer area. Aquifer saturation depth in each simulation month is calculated from the difference between average level in piezometers and aquifer average depth for the same month. Then, the calculated saturated depths are compared with the observed ones. The average of variance between them was 0.61 (Table 9 and Fig. 3).

MODFLOW model

Balance factors used in the MODFLOW model for the Tuyserkan aquifer are as follows:

(1) Input factors:

Input groundwater flow
Rainfall infiltration
Surface flow infiltration
Return flow from agricultural usage
Return flow from municipal usage

Table 9 Calculated and observed saturated depths after WEAP calibration

Month	Calculated saturated depths (m)	Observed saturated depths (m)
1	23.18	74.17
2	6.18	44.18
3	1.19	21.19
4	17.19	29.20
5	03.19	76.20
6	73.18	94.19
7	33.18	68.18
8	55.17	32.17
9	03.17	46.16
10	58.16	9.15
11	8.16	11.16
12	55.16	51.16

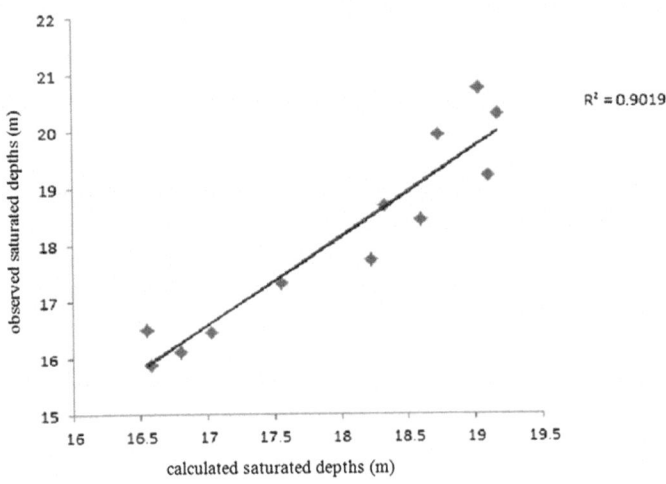

Fig. 3 Observed and calculated saturated depths after WEAP calibration

(2) Output factors:

Output groundwater flow

Groundwater discharge by wells

In order to grid the Tuyserkan aquifer, cells on a 500 × 500 meters scale were applied. The result was a mesh consisting of 3564 cells (66 × 54). Due to the aquifer's condition (lack of lakes or rivers with constant head) there are no cells with constant head. The model consists of a mesh of active and inactive cells. Simulated boundaries in this model include the No Flow Boundary and General-Head Boundary. Kriging was used to calculate the aquifer density.

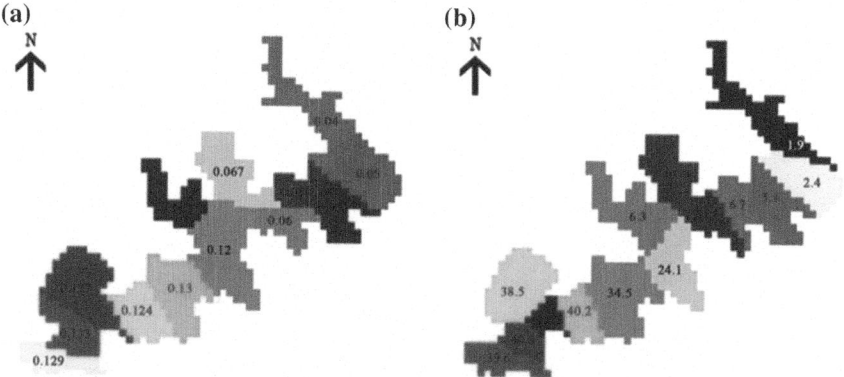

Fig. 4 Hydraulic conductivity (**a**) and storage coefficient values after MODFLOW calibration

Considering the Kriging processes' dependency on geostatistics, for each set of data (i.e. bedrock elevation, ground surface elevation and groundwater elevation) a Gaussian model was chosen by GS+ software and then interred into the Field Interpolator device in the PMWIN software. By considering factors such as crop pattern, well discharge and rainfall conditions, 12 stress periods were considered for 2009–2010. Thirteen (13) piezometers were used for simulating the aquifer in MODFLOW (Fig. 2). Hydraulic conductivity and storage coefficient parameters were used to calibrate the model in 22 December 2009 to 20 January 2010. Also recharge parameter calibrated for stress period. Sensitivity analysis was done for hydraulic conductivity, storage coefficient and recharge rate parameters. Calibration

Fig. 5 Observed and calculated groundwater levels after recharge calibration

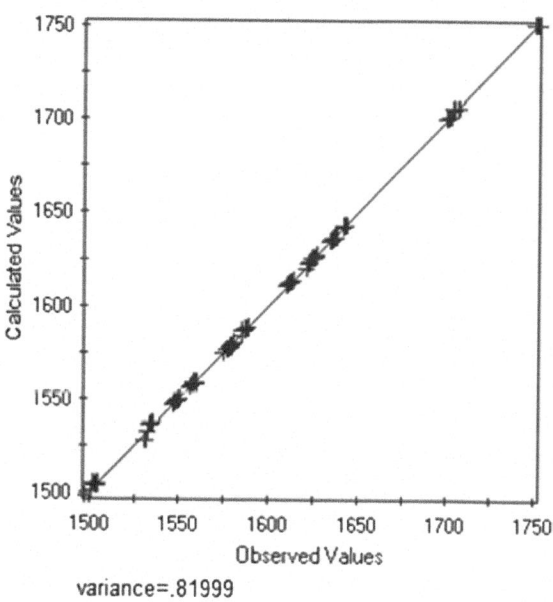

variance=.81999

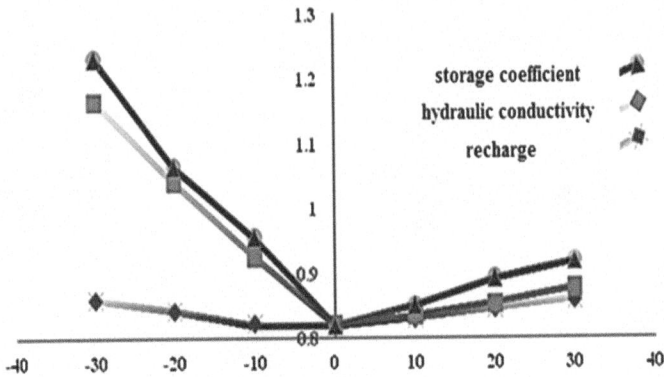

Fig. 6 Sensitivity analysis results

results in steady and transient states indicate that, horizontal hydraulic conductivity (Fig. 4a) and storage coefficient (Fig. 4b) increases from north east to the south west in the Toyserkan plain. Recharge rate calibration results indicate that error variance for model is 0.82 (Fig. 5). Sensitivity analyses demonstrated that the water levels in the Tuyserkan aquifer were most sensitive to the storage coefficient, the hydraulic conductivity and the recharge rate (Fig. 6).

Reference scenario

In this scenario, first agricultural, industrial and municipal water requirements of the reference year (2011 for WEAP and 2009 for MODFLOW) were applied to the models. Municipal water requirements of urban and rural areas in the region were applied by considering the population growth rate, but the potential agricultural and industrial water requirements were assumed as stable. Recharge was determined by estimated rainfall average. The predictions provided by the WEAP and MODFLOW models are to 2025 and 2019, respectively

Efficiency scenario

The efficiency scenario was applied under the assumption that all land covered by groundwater resources is irrigated by pressurized systems, thus resulting in a 20% increase in efficiency.

3   Discussion

WEAP model results (reference scenario)

After applying this scenario, the results demonstrated that water supply will face a severe crisis to 2019. These changes are shown in Figs. 7 and 8.

In this scenario, the entire plain's water requirement is not secured. In the year 2020, the overall water requirement of agricultural sites will not be covered. The Ghogholrood and Mianrood sites, located in the aquifer downstream, will not be

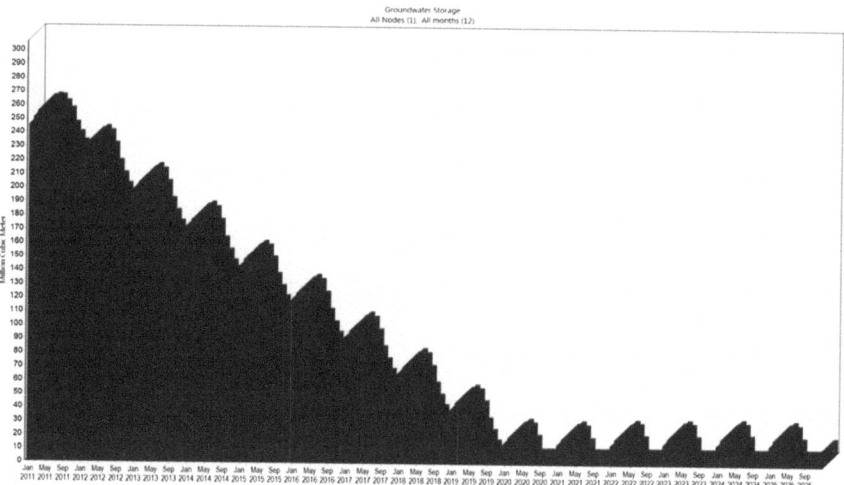

Fig. 7 Reservoir aquifer volume change in the simulation period

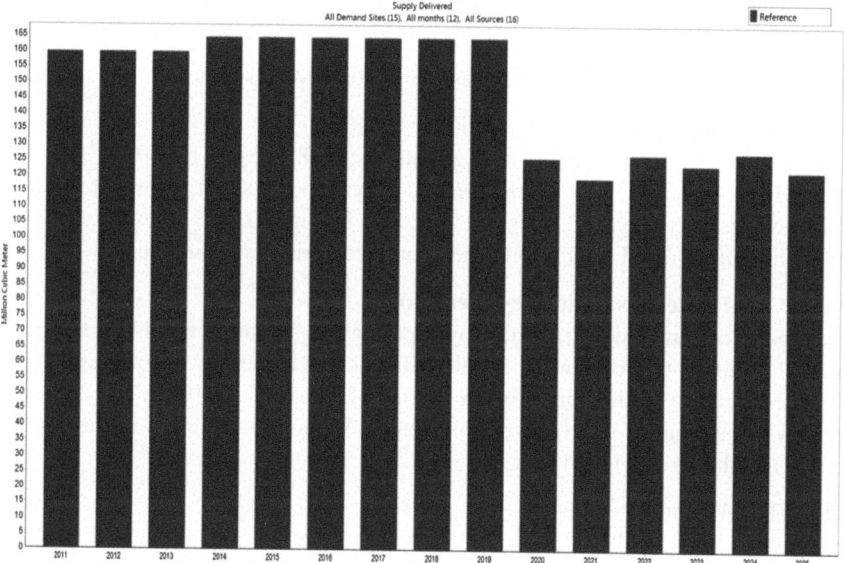

Fig. 8 Water supply to the all demand sites in the simulation period

able to secure their cultivation requirements in the months following summer 2020 (Fig. 9).

However, in the simulation period (25 years), the Korzanrood and Hayaghoogh Nabi sites (located in the aquifer upstream) faced almost no droughts. The minimum of monthly requirement in these sites will be 84.5%. The highest unsecured

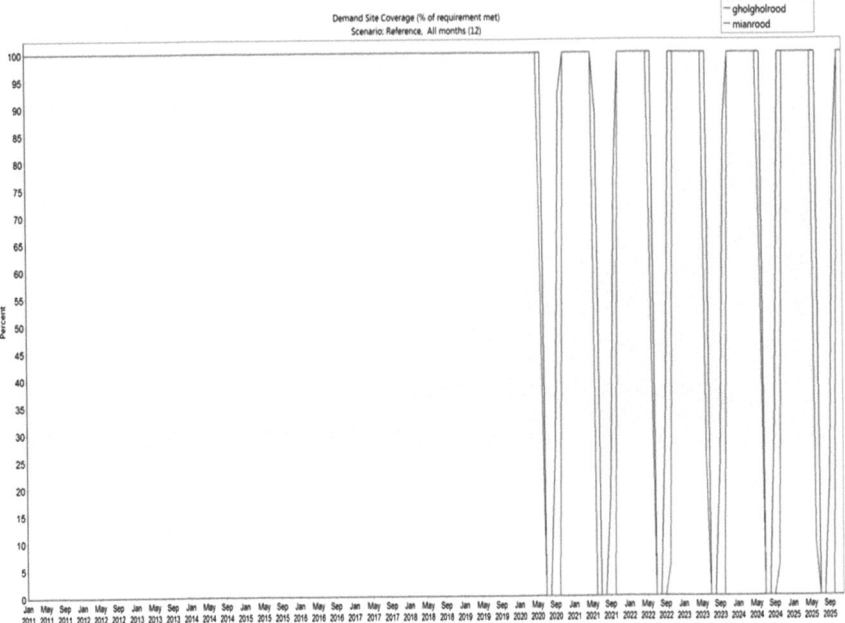

Fig. 9 Percent of supply requirements in Gholgholrood and Mianrood sites

requirement for cultivation sites were as follows: 20.8 10^6 m^3, 1.8 10^6 m^3, 1.1 10^6 m^3 and 20.7 10^6 m^3 for Ghogholrood, Hayaghoogh Nabi, Korzanrood and Mianrood, respectively. All occurred in 2021. In Fig. 10, the amount of annual input and output of all sites during the simulated period is presented.

WEAP model Results (efficiency scenario)
The models output demonstrates that by applying this scenario, not only is the crisis in 2020 avoided but groundwater supply will reach its highest level compared to the reference scenario, with an increase in volume of up to 287.62 10^6 m^3. In 20 years, the Tuyserkan aquifer will recover completely. Volumetric changes in the aquifer supply are shown in Fig. 11.

MODFLOW model results
The results from applying the data from the reference scenario and efficiency scenario to the MODFLOW model are shown in Figs. 12 and 13.

According to the estimated hydrographs in both scenarios, except for two piezometers (3 and 6), in the other piezometers, the groundwater level at the end of simulated period declined compared to its initial level. In the first scenario, piezometer 8 has the highest decrease, equal to 25.83 m and piezometer 5 has the lowest (2.285 m). Groundwater level increase in piezometer 3 is 9.022 m and in piezometer 6 is 15.066 m. The highest decrease in the second scenario is also recorded to piezometer 8, although the decrease is lowered to 4.531 m and the groundwater level is 21.3 m. Piezometer 9 has the lowest decrease (1.352 m). The

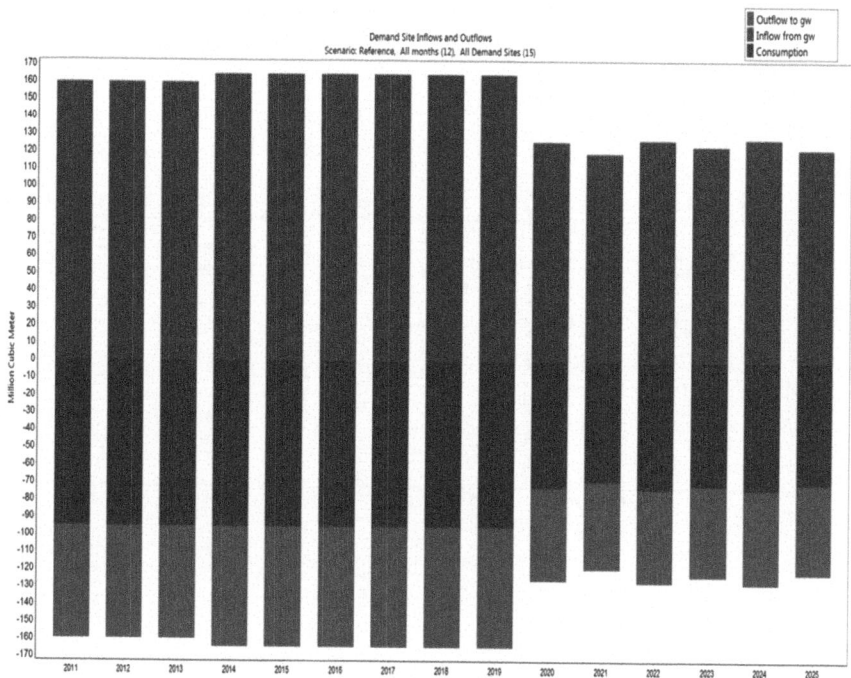

Fig. 10 The annual amount of inflow and outflow of all demand sites in simulation period

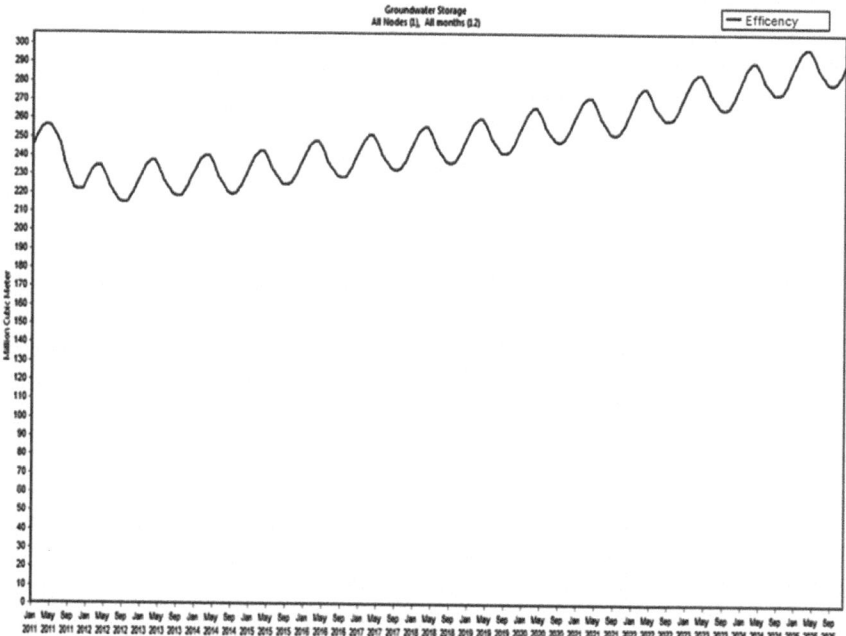

Fig. 11 Reservoir aquifer volume change in the simulation period

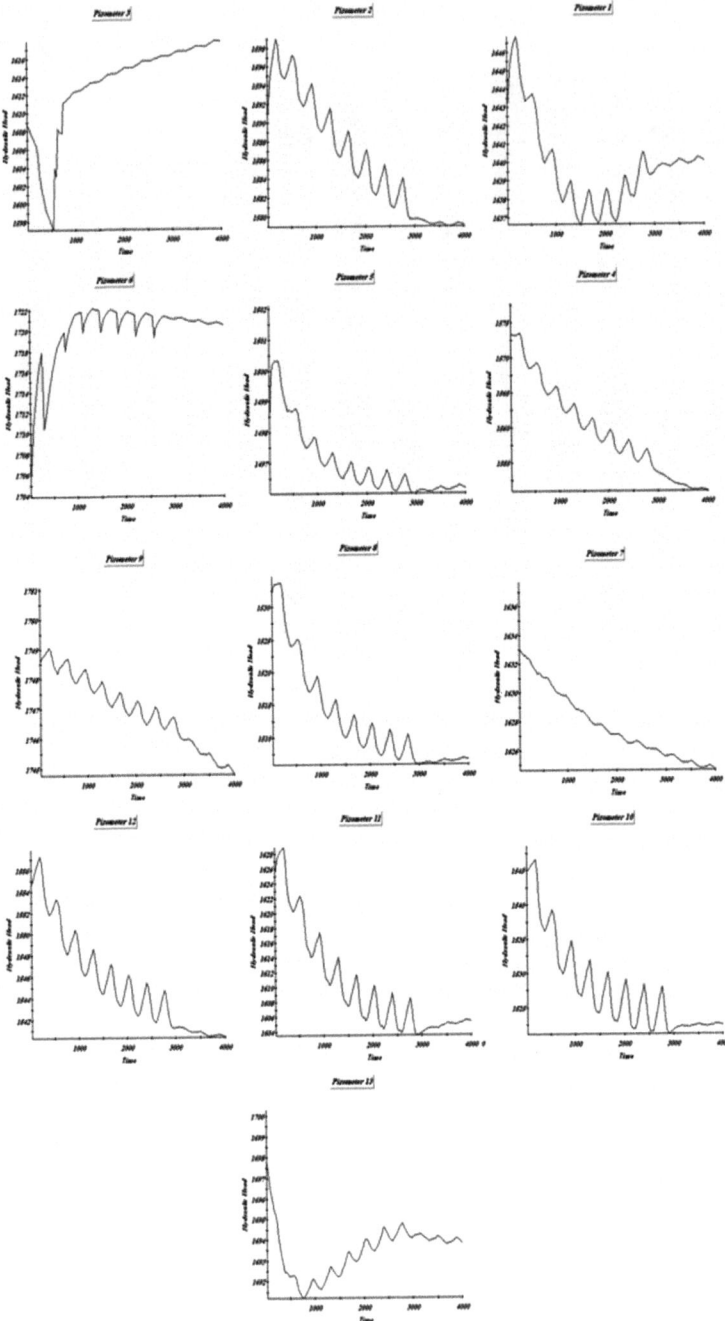

Fig. 12 MODFLOW output after running the reference scenario

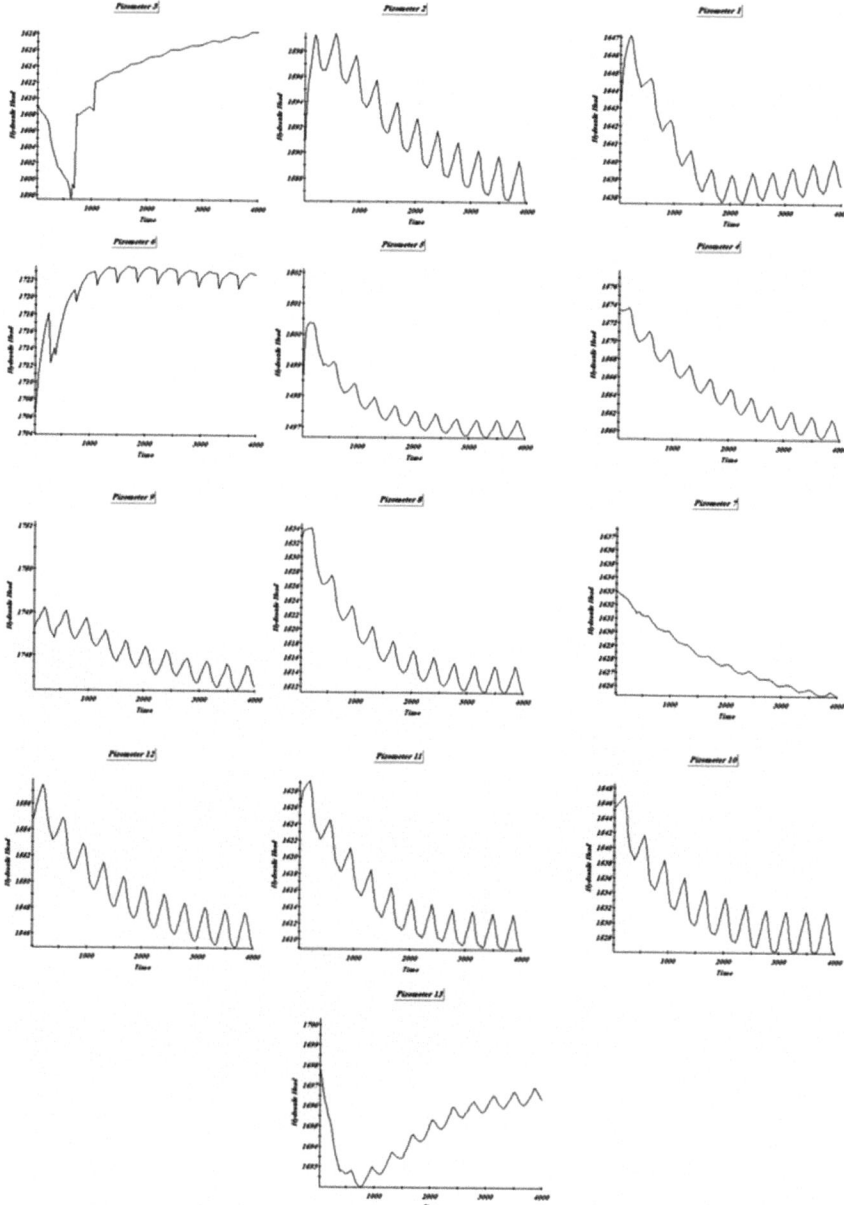

Fig. 13 MODFLOW output after running the efficiency scenario

lowest difference is recorded to piezometer 5 (0.516 m). At the end of the simulated period, the groundwater level in piezometer 3 has a 9.264 m increase from its initial condition which, compared to the first scenario, only shows a 0.242 m difference.

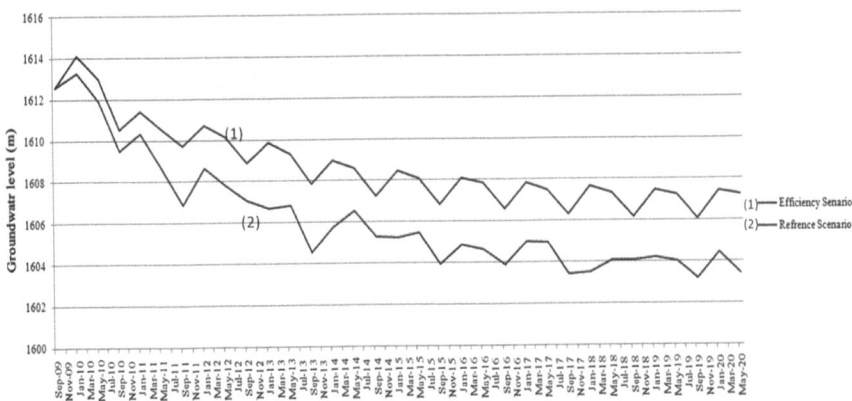

Fig. 14 Compare the average annual groundwater level for two scenarios in Tuyserkan aquifer (MODFLOW output)

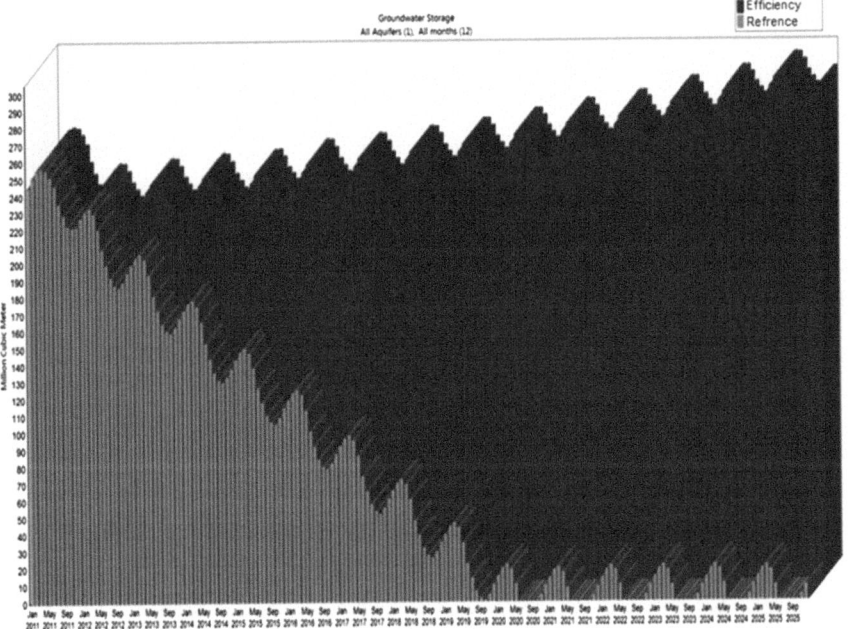

Fig. 15 Compare the reservoir aquifer volume change in two scenarios (WEAP output)

However, in the case of piezometer 6, applying the second scenario caused its groundwater level to increase by 1.661 m and reach 16.727 m. In applying the changes in hydrographs, a comparison was made between estimated rainfall and discharge in both scenarios for each particular piezometer. These comparisons show that all of the hydrographs corresponded to rainfall estimations. However, in some of the piezometers such as number 6, there seems to be a slight difference in estimated groundwater level due to using the linear solution method. Piezometer 6 is located in a position that changes in height over a short spatial distance; in other words, it has a high slope.

Comparing two scenarios in WEAP and MODFLOW models

In the MODFLOW model, after determining the average annual groundwater level throughout the aquifer, an attempt was made to draw an annual comparison between the two models. As shown in Fig. 14, although average groundwater level in both scenarios declines, the average groundwater level at the end of the simulated ten-year period in the first scenario is 3 m lower compared to the second scenario.

Also, the 20% increase in irrigation efficiency caused the groundwater level to be more stable in the resulting hydrograph. Figure 15 shows the reservoir aquifer volume change in efficiency and reference scenarios in the WEAP model.

4   Conclusions

According to simulation of the aquifer in the WEAP model, in the reference scenario the Ghogholrood and Mianrood sites will face a serious crisis in the upcoming years. The results from the MODFLOW model also confirm this process. Seven piezometer hydrographs (2, 4, 5, 7, 8, 10 and 12), which are located in these two agricultural sites, will considerably decline in the next few years. However, the piezometers located at the Korzanrood and Hayaghoogh Nabi sites (1, 3, 6, 9, 11 and 13) will have a much lower decrease and in some places, the groundwater level will increase in the next few years. According to the WEAP model results, these agricultural sites will have the lowest amount of water demand and almost no drought problems.

According to the results from the efficiency scenario in both the WEAP and MODFLOW models, the groundwater level will increase in the next few years. It should however be mentioned that the two models have different starting and ending points in hydrographs and charts. The common sections in both models have an approximately similar process, and results from both models show an increase in groundwater level under the efficiency scenario.

It should be noted that these scenarios are defined by considering the current crop pattern, similar rainfall and discharge trend to that in the next year. Stopping crops with high water requirements (such as clover, alfalfa, sugar beet, tomato, cucumber and beans), and using crops with low water requirements, in addition to increasing irrigation efficiency, will increase the aquifer supply.

References

Ahadi, R., Samani, Z., & Skaggs, R. (2013). Evaluating on-farm irrigation efficiency across the watershed: A case study of New Mexico's Lower Rio Grande Basin. *Agricultural Water Management, 124,* 52–57.

Alizadeh, H. A., Liaghat, A., & Sohrabi, T. (2014). Assessing pressurized irrigation systems development scenarios on groundwater resources using system dynamics modeling. *Journal of water and soil resources conservation, 3*(4), 1–15.

Bournaris, T., Papathanasiou, J., Manos, B., Kazakis, N., & Voudouris, K. (2015). Support of irrigation water use and ecofriendly decision process in agricultural production planning. *Operational Research: An International Journal, 15,* 289–306.

Burt, M., Clemmens, A. J., Strelkoff, T. S., Solomon, K. H., Bliesner, R. D., Hardy, L. A., et al. (1997). Irrigation performance measures: Efficiency and uniformity. *Journal of Irrigation and Drainage Engineering, 123*(6), 423–442.

Cooley, H., Christian-Smith, J., & Gleick, P. (2009). *Sustaining California agriculture in an uncertain future.* Pacific Institute: Technical Report.

Gleeson, T., Wada, Y., Bierkens, M. F., & van Beek, L. P. (2012). Water balance of global aquifers revealed by groundwater footprint. *Nature, 488*(7410), 197–200.

IMPO Water Affairs Bureau (Management and Planning Organisation I. R. Iran). (2003). *Report of water sectors 'operation and problems' analysis* (in Persian).

Jafary, F. (2016). *Participatory modelling platform for groundwater irrigation management with local farmers in Iran (Kashan).* Thesis for the degree of Doctor of Philosophy, School of Geography, Earth and Environmental Sciences, University of Birmingham, 359 p.

Mahdavi, M., Farrokhzade, B., Salajeghe, A., Malakian, A., & Souri, M. (2012). Simulation of Hamedan-Bahar aquifer and investigation of management scenarios by using PMWIN. *Watershed Management Research (Pajouhesh & Sazandegi), 98,* 108–116.

Manos, B., Bournaris, Th, Papathansiou, J., & Voudouris, K. (2009). A Decision Support System (DSS) for sustainable development and environmental protection of agricultural regions. *Environmental Monitoring and Assessment, 164*(1), 43–52.

Mehta, V. K., Haden, V. R., Joyce, B. A., Purkey, D. R., & Jackson, L. E. (2013). Irrigation demand and supply, given projections of climate and land-use change in Yolo County, California. *Agricultural Water Management, 117,* 70–82.

Pfeiffer, L., & Lin, C. Y. C. (2014). Does efficient irrigation technology lead to reduced groundwater extraction? Empirical evidence. *Journal of Environmental Economics and Management, 67,* 189–208.

Regional Water Organization of Hamadan. (2008). Water resources in Hamadan area. Ministry of Energy, Iran Water Resources Management

Seiber, J., Swartz, C., & Huber-Lee, A. (2005). User guide for WEAP21, Stockholm Environment Institute Tellus Institute, 176 p.p.

Shiklomanov, I. A., & Rodda J. C. (2003). *World water resources at the beginning of the twenty-first century.* Cambridge, UK, Cambridge University Press.

Singh, A. (2014). Conjunctive use of water resources for sustainable irrigated agriculture. *Journal of Hydrology, 519,* 1688–1697.

Sulis, A., & Sechi, G. M. (2013). Comparison of generic simulation models for water resource systems. *Environmental Modelling and Software, 40,* 214–225.

Taheri Tizro, A., Voudouris, K. S., & Akbari, K. (2011). Simulation of a groundwater artificial recharge in a semi-arid region of Iran. *Irrigation and Drainage, 60,* 393–403.

Yazdanpanah, T., Khodashenas, S. R., Davari, K., & Ghahraman, B. (2008). Water resources management in a Watershed with WEAP model (A case study of Azghad watershed). *Journal of Agriculture Science and Technology, 22*(1), 213–221.
Zadevakili, N. (2011). *Surface and groundwater resource allocation policies using the integrated operation model* (*A case study of Zayanderood basin*). Thesis submitted for Master of Science, Isfahan University of Technology, Department of Civil Engineering.
Zhou, Y., & Li, W. (2011). A review of regional groundwater flow modeling. *Geosciences Frontiers, 2*(2), 205–214.

The manufacturer's authorised representative in the EU is Springer
Nature Customer Service Centre GmbH, Europaplatz 3, 69115 Heidelberg,
Germany. If you have any concerns regarding our products, please
contact ProductSafety@springernature.com

Printed and bound by CPI Group (UK) Ltd, Croydon, CR0 4YY

29/04/2026

02099458-0018